Purchasing in
the 21st Century

Purchasing in the 21st Century

A Guide to State-of-the-Art Techniques and Strategies

John E. Schorr

THE OLIVER WIGHT COMPANIES

Oliver Wight Publications, Inc.
5 Oliver Wight Drive
Essex Junction, Vermont 05452-9985

Library of Congress Catalog Card Number: 92-060538

ISBN: 0-939246-22-8

Printed on acid-free paper.

Manufactured in the United States of America.

10   9   8   7   6   5   4   3   2   1

Contents

Preface

This book contains practical, how-to information about implementing supplier scheduling and carrying out purchasing in a Manufacturing Resource Planning (MRP II), Just-in-Time (JIT), and Total Quality Control (TQC) environment. It not only provides insights into the basic tools available to the purchasing professional, but focuses on the importance of developing valid schedules and continually improving purchasing performance using MRP II/JIT/TQC.

The introduction sets the tone for purchasing as we move into the 21st century. Chapter 1 explores the benefits of using supplier scheduling, MRP II, JIT, and TQC. Chapters 2 and 10 discuss what is required of both sides, the supplier and the buyer, in a partnership environment. Chapter 3 describes the basics of supplier scheduling under MRP II, while Chapter 4 explains how the JIT/TQC process applies to purchasing.

After explaining the need for valid schedules and the JIT/TQC concept, the balance of the book covers the specific tools and techniques used to improve performance. Chapter 5 covers the organization needed for supplier scheduling while Chapter 6 covers the need for supplier quality assurance. Chapter 7 discusses the difference between single-sourcing and sole-sourcing, as well as the reasons for reducing the supplier base. Chapter 8 discusses the need to synchronize with your suppliers' production schedules. It also includes techniques to reduce the freight costs on purchased components in a JIT environment when the buyer is getting smaller, more frequent shipments.

Chapter 9 describes the need for partnerships with suppliers and suggests a set of "Golden Rules." The chapter also covers the concepts of supply chain management and the use of electronic data exchange in purchasing. Chapter 11 covers special situations, such as buying commodities in short supply.

Performance measurements and management information necessary to continuously improve performance, the implementation of supplier scheduling and MRP II/JIT/TQC in purchasing, and supplier education are covered in Chapters 12, 13, and 14. The fifteenth and final chapter reinforces the need to create a spark, ignite change, and use the one-less-at-a-time JIT approach.

Introduction

My career in purchasing began in 1965, when I accepted my first job as a junior buyer for a large company. I was responsible for buying several of the items used in the company's Manufacturing Department. During my first week on the job, I was directed to read the Purchasing Policy and Procedures Manual and learn "how to buy." I quickly realized, though, that the buyers around me were not going by the book. The manual did not explain how to expedite, yet expediting seemed to be the normal way of doing business. So for the rest of the 1960s, I learned that the informal system really ran the company. We used "Red Tags" to move the hot orders through the plant and we expedited all day long. When we ordered items, we always were required to buy an economic lot size although the unstated rules said, "Always have some on order and never run out of anything." Not surprisingly, we felt we were understaffed and management just didn't understand our problems.

Then the 1970s came along and I got my first exposure to the use of computers in purchasing. The computer, plus an MRP system, changed my life. The computer always seemed to "screw up" and the inventory on the computer was never right. The company started talking about a formal system, but I still spent most of the day expediting. While the environment changed, the only thing that seemed formalized to me was the shortage list. I don't remember having a shortage list in the 1960s; the planner would bring the traveling requisition or cardex card to me and I would check the status of the part. The computer, which was used for inventory planning, gave us the formal shortage list.

The 1980s saw offshore competitors taking large segments of our markets and causing severe profit pressures. All of a sudden, purchasing professionals had a new set of terms to deal with. These included Manufacturing Resource Planning, Just-in-Time, Total Quality Control, Electronic Data Interchange, closed loop systems, performance measurement, and accountability.

And as we entered the 1990s, yet another new set of terms began to appear: Continuous Improvement, Supplier Partnerships, Customer/ Supplier Linkages, and Benchmarking, to name a few.

Despite the dramatic changes that purchasing has undergone in the more than twenty-five years since I entered the profession, the typical buyer is still buried in paperwork and caught up in expediting, still using outmoded systems, and still has little opportunity to do the job better.

This book is designed to help the purchasing professional spearhead an effort for getting out from under the inefficiencies of the informal system and realizing maximum benefit from the formal tools available today. A 1988 Oliver Wight survey of more than a thousand companies showed how large these benefits can be in purchasing. Those companies that had a "Class A" MRP II/JIT/TQC system enjoyed an average 96 percent on-time and a *13 percent* annual cost reduction in purchasing! Imagine what a 13 percent purchased cost reduction would do for your company. Helping you achieve such a reduction is what this book is all about.

Purchasing in
the 21st Century

Chapter One

Purchasing's New Leadership Role

THE HEAT IS ON

Purchasing professionals in companies that lack Manufacturing Resource Planning (MRP II), Just-in-Time (JIT), and Total Quality Control (TQC) may question the value of such tools. They typically ask, "How will they help me do a better job of buying?" The answer is very simple: The tools give them the time and information they need to focus on things that actually save their company money.

Today, because of competitive pressures from the Pacific Rim, manufacturing companies are learning to produce more product with fewer people. In other words, companies are doing more and more purchasing and adding less value internally. This has caused the ratio of purchased dollars to direct labor dollars in the typical manufacturing company to change from a three-to-one ratio in the mid-1980s to a four-to-one ratio in the early 1990s. In other words, for every one dollar spent on direct labor, the company is spending four dollars on direct purchased material. Most companies have a staff of skilled engineers who develop better methods in manufacturing and continue to reduce the labor content. As they succeed, their efforts will drive up the ratio to five to one or six to one by the 21st century. As the direct labor content continues to decline, purchasing professionals will come under greater pressure to find ways of cutting the costs of direct purchased materials.

The traditional way for Purchasing to reduce costs is to buy in larger order quantities. Not only are the prices associated with larger quantities

3

lower, but often so are the freight costs and paperwork costs. But that is exactly the opposite of what the buyer is asked to do in a Just-in-Time environment; one of the goals of Just-in-Time is to drive the inventory down dramatically. Therefore, the buyer will be expected to bring in a high percentage of the items in smaller lot sizes, more frequently delivered. Instead of bringing in items on a monthly basis, weekly receipts will be expected. And in companies used to weekly receipts, daily receipts will be requested. The buyer, therefore, will not only be expected to lower the costs, but he or she will have to do so with smaller lot sizes.

As the company becomes more involved with Just-in-Time and Total Quality Control and starts bringing in items when they are needed, buyers will be under tremendous pressure to get defect-free, quality goods from their suppliers. The reasoning is simple: With little or no inventory on hand, the production line will have to shut down if incoming parts are rejected.

Also, to respond better to the ever-changing needs of customers, Purchasing will be requested to negotiate shorter, more stable lead times with their suppliers. With little or no inventory on the items, the suppliers will have to be able to respond quicker to changes in the marketplace.

Finally, smaller, more frequent receipts from suppliers increases the pressure to manage freight costs. Unless the buyer is able to effectively control inbound freight costs, the smaller lot sizes could drive up transportation costs and offset any savings associated with the inventory reduction. The buyer will therefore be expected to find creative ways to lower the freight rates.

To recap, the heat will be on Purchasing to reduce costs and increase flexibility while getting smaller lot sizes, more frequent shipments, defect-free quality, shorter, more stable lead times, and less inventory.

The problem buyers experience in the traditional setting is that they are too busy expediting and scaling mountains of paperwork to do an effective job. Even if they could find the time, they typically don't have all the information they need to negotiate the best deal for their company. The net result is that significant opportunities for savings are missed, which in turn means that companies aren't as competitive as they might be and purchasing people find themselves under constant pressure to reduce costs. The good news, though, is that all the tools necessary for creating and taking advantage of cost-saving opportunities, with the MRP II and JIT/TQC plans, are available today.

MRP II is a tried and proven planning and scheduling tool. It provides a rack-and-pinion connection between high-level and detail plans, so that as the business plans change to meet the demands of the marketplace, the detail plans can be changed and executed as well. MRP II closely integrates the purchasing function to all parts of the business, which means that the purchasing professional can provide suppliers with solid schedules of the company's requirements.

JIT is not a tool for minimizing inventory or delivering parts when they're needed, as is commonly believed; rather, it is a methodology that leads to continual improvement by eliminating waste in Manufacturing, Purchasing, and all other departments within the company. As we'll see throughout this book, JIT coupled with TQC has enormous impact on the purchasing process, because it drastically reduces the number of suppliers a company deals with, eliminates a great deal of paperwork, and drives the continuous improvement process toward shorter, more stable lead times, more frequent deliveries, better quality, on-time delivery, and lower costs.

TQC—Total Quality Control—is the flip side of Just-in-Time. TQC is a collection of tools for eliminating defects from design engineering through delivery of the finished product. While JIT is a methodology for eliminating waste and uncovering problems, TQC is a set of tools for solving those problems and making continual improvements. Like JIT, TQC has major implications for the purchasing professional, such as source assurance of quality, which eliminates paperwork, simplifies communication, and makes it possible to eliminate the need for separate inspections of incoming materials.

In the following chapters, we'll describe each tool in greater detail. First, though, let's imagine what life would be like if your company adopted the winning combination of MRP II and JIT/TQC.

PARADISE FOUND

To appreciate the way things could be in your company, we'll first review the traditional way Purchasing interacts with the rest of the company and its outside suppliers:

Round and Round the Merry-Go-Round

1. The material planner determines the need to buy more materials.

2. Next, the planner sends the buyer a written requisition for the materials to be purchased.

3. Upon receipt of the requisition, the buyer prepares a PO and shoots it off to the appropriate supplier.

4. If the supplier company can fill the order, it sends back an acknowledgment. If not, it typically suggests alternative quantities or delivery dates.

5. If the supplier acknowledges that the order can be filled, the transaction proceeds along a normal course. If the supplier company has proposed an alternative, the buyer will have to check with the planner and/or the engineer to see if the alternative fits in with the production schedule. If not, the buyer might then have to recontact the supplier and work out a compromise. Even after a compromise is found, changes in production schedules may alter the requirements for the purchased material. In that case, the planner must again contact the buyer, who must again contact the supplier, and the merry-go-round continues.

The Major Drawbacks

Many purchasing veterans will tell you that the traditional way of buying materials is fraught with problems, the main ones being a lack of valid schedules, poor communication, too much paperwork, too much wasted time and energy, conflicting directions, and teamwork problems. Let's take a closer look at each of these major problem areas.

The first major problem area is the lack of valid schedules. The dates that Purchasing gets on the requisitions are not need dates—when the company needs the first piece for production—but rather replenishment dates—when the company would like to have more product in inventory. Because the company experiences many changes in the schedules due to inaccurate forecasts, customer changes, scrap and rework, engineering changes, new equipment, equipment breakdowns, etc., it may want to carry safety stock. For example, the company might choose to have one or two months' supply of purchase material inventory on hand at all times to buffer against production schedule changes. Purchasing in these companies might be charged with buying material to maintain an inventory level, not to support a particular production schedule. If the due date on the requisition is from four to eight weeks before the need date, the buyer knows that the material is being ordered to maintain the safety stock, and so does the supplier. (The situation becomes evident if there are purchase orders two or three weeks past due and the company hasn't run out of the material yet.) As a result, if a problem occurs at the supplier's plant, the buyer typically has no way of knowing exactly when the items are really needed.

The second problem area is communication. The traditional method of purchasing requires intensive communication within the company and between the company and its suppliers. Communication is also extremely inefficient, because no one has the "big picture." Only the planner knows what the company needs, and only the supplier's people know what their company can deliver. The buyer functions as a middle-man, relaying information between the two. Unfortunately, information conveyed in this manner is always subject to miscommunication and interpretation. And once the communication is out of sync, the result may be inaction, with the attendant lost time, or a decision by default, which may not turn out to be a very good decision at all. In addition, every buyer has, at one time or another, wasted a fair amount of time and energy just playing telephone tag with the supplier's personnel in trying to relay information.

The third major problem area is paperwork. The merry-go-round between the company and supplier runs on paper. And lots of it. Purchasing Departments spend thousands of hours just processing requisitions, reviewing and signing PO's, matching up acknowledgments with PO's, generating change orders, matching change orders with original orders, and carrying out other paper-intensive tasks.

The fourth major problem area is wasted time and energy. As we stated earlier, the dates on purchase orders aren't valid. In fact, the dates often change before the shipment is due. The buyer has some orders due in two weeks that they are being asked to bring in ASAP. The buyer also has some orders that are two weeks past due and no one is asking for them. As a result, buyers must spend much of their time expediting all orders just to keep the production line running. Otherwise, people and equipment would be sitting idle while they waited for material to arrive. And yet expediting is also an extremely inefficient way of doing business; it not only leads to higher costs and diminished customer service, but diminishes the overall quality of life in the workplace. Typical purchasing personnel, when asked if they would like to be hired or fired based on 95 percent on-time to the dates they get on the requisitions, say, "No way."

One other major problem area needs to be mentioned: the conflicting directions that the buyer receives from management. The buyer's job description cites controlling costs and cost reductions as a major responsibility, but all the hidden messages buyers receive paint quite another picture. Every time buyers save a dollar, they get an "Attaboy or Attagirl." Every time buyers are short material and cause a production line to go down, they get a "Dummy." It takes 863 "Attaboys" to equal one "Dummy," so the buyer quickly learns to work to the shortage list first and tend to cost reduction second.

Unfortunately, when management sends the auditors into Purchasing to determine how well Purchasing is performing, they spend their time seeing if all the pieces of paper were handled correctly and had the correct approvals, not how well the department performed in the areas of cost reduction, delivery, and quality. So the buyer now learns that keeping the production lines running is the first priority, handling all the paperwork correctly is the second priority, and doing cost reduction is the third priority—if time is available.

Next, management gets concerned about the inventory being too high and has a discussion with the Inventory Control Department about reducing inventory levels. As a result, Inventory Control reduces the order quantities on the requisitions it sends to Purchasing. This just makes it all the more difficult for Purchasing to control costs. The buyer goes to Inventory Control and requests an increase in the order quantity and is told "No, we need to control inventory levels." The buyer's job description may say cost reduction, but the hidden messages say something else.

Therefore, the typical buyer's day looks like the pie chart in Figure 1.1. As you can see, most of the day is spent expediting, doing paperwork, and attending meetings, leaving very little time to spend money well.

Other Drawbacks

In addition to the problems cited above with the flow of information and conflicting directions, the traditional approach to purchasing carries with it several subtle, but nevertheless serious, drawbacks.

The traditional approach perpetuates long lead times. Long lead times are a company's natural enemies. They're bad for a company because it appears less flexible; they make it difficult to phase in new product; they tie up cash in inventories; and they subject the production process to changes and disruptions because they force people to commit further into the future and guess more about what the company needs.

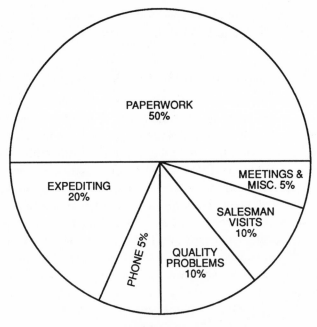

Figure 1.1

Lead times quoted by suppliers can often expand into the hazy future. Unfortunately, with the traditional way of purchasing materials, you're at the mercy of the supplier's current business level or backlog. Since suppliers typically have little forward-planning information other than the current customer orders, they often place new incoming orders at the end of the line.

The traditional approach also perpetuates teamwork and accountability problems. As mentioned earlier, purchasing people are caught between a rock and a hard place. On the one hand, management directs them to order in very small quantities to keep the inventories down. On the other hand, the suppliers require large order quantities before they will give attractive price breaks. The net result is a tug-of-war between Purchasing and Production Control, so that the two departments compete against each other rather than the real "enemies" down the street and across the ocean.

Because of all of the above-mentioned problems, the traditional way of doing purchasing is replete with waste and opportunities for errors. And when people have to deal with a system that they can't trust, they're generally not eager to put their jobs and careers on the line. In other words, they won't assume accountability for their decisions and actions. And without accountability, no system can operate effectively.

Operating Under the MRP II/JIT/TQC Umbrella

In contrast to the traditional approach, consider what life is like when a company operates with MRP II/JIT/TQC tools. Companies such as Xerox and Black & Decker have long-term supplier contracts that give them a stable price over the life of the contract but allow the company to buy only the quantity needed to support their current production schedule. The schedules both the buyer and suppliers work to are valid and represent the real need dates for the material. Routinely—at least weekly—schedules of the company's requirements are transmitted to the suppliers, ideally electronically. The suppliers produce in sync with the company's own internal production facilities. They frequently make deliveries to a company once a week, once a day, or several times a day, depending on the nature of the company's business. The suppliers assure the quality of the parts, so as a result, there's no need for incoming inspections, and parts can be delivered by the supplier directly to the point-of-use at the buyer's company. Because this cycle repeats itself routinely, like clockwork, the buyer now finds the time to do what his or her job description states—getting the materials on time, in specification, at the lowest total cost for all the items purchased.

The company's customers are also tied into the MRP II network in a supply pipeline arrangement that allows everyone—the customer, the production facility, and the supplier—to synchronize production schedules, to produce only what is needed when it is needed, and to eliminate excess inventories, poor communication, and excess costs.

The typical buyer's day is now dramatically different from that represented by the pie chart shown in Figure 1.1. As Figure 1.2 shows, fully 50 percent of the buyer's time is now spent on cost reduction and negotiation. Thirty-five percent of the time is spent gaining a better understanding of the requirements of each part and how it is used, as well as a better understanding of how those requirements can be met in shorter lead times, smaller lot sizes, and at the lowest possible cost. Only 15 percent of the buyer's time is spent expediting, but the expediting occurs before the item hits the shortage list. Because the buyer has a clear picture of future requirements and the MRP system nets the on-hand and the on-order quantities against inventories, the buyer can see

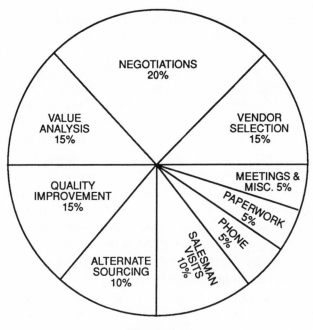

Figure 1.2

potential shortages in advance. This gives the buyer the opportunity to prevent the shortage from occurring in the first place.

The Major Benefits

The MRP II/JIT/TQC umbrella offers many advantages over the traditional merry-go-round approaches. The two biggest advantages are in the areas of valid schedules and cost reduction/value analysis. As mentioned in the Introduction, the Class A users of MRP II/JIT/TQC reported on average 96 percent on-time and a 13 percent cost reduction in the 1988 survey. In a company that has six dollars in purchasing for every one dollar in direct labor, that 13 percent savings is equivalent to the entire direct labor cost for the year. And a dollar saved in purchasing goes directly to the bottom line of a company's balance sheet.

Of the many benefits of doing purchasing in the style described in this book, here are the key ones that any company will appreciate. To begin with, your supplier can deliver the right material at just the right time, on a routine basis. The supplier's shipping schedules and internal production schedules will also be in sync, so both of you are producing at the same rates. In addition, you can bring in quantities that exactly match production requirements, without being penalized for buying in small quantities.

Next, you deal with fewer suppliers, which means that you develop long-term "partnerships" rather than short-term adversarial relationships. When the supplier ships the material, you will be so confident of its quality that it can go straight to the production floor. Purchased-material inventories can be dramatically cut, while actually improving service to manufacturing. You can also operate with supplier lead times of a few days or weeks rather than many months.

Then there's the issue of teamwork—people can finally work together as a team and legitimately be held accountable for doing their jobs. Purchasing professionals will have the time to do the really important parts of their jobs, too: sourcing, negotiating, contracting, cost reduction, value analysis, supplier certification, and other actions that help drive down the cost of purchase materials.

Finally, long-term partnerships with carriers enable them to make more frequent deliveries to you without charging premium freight charges.

BECOME A CATALYST FOR CHANGE

If the benefits of operating in an MRP II/JIT/TQC environment sound like a wish list, be assured that they are being realized at this very moment by companies throughout the world, very possibly by your own competitors. If your company is not presently using MRP II/JIT/TQC to its full potential, your mandate as a purchasing professional is to act as a catalyst and demonstrate the invaluable contribution these programs can make to the company's bottom line. No doubt you'll encounter resistance; after all, the concept of shrinking the supplier base runs counter to traditional wisdom, especially in purchasing, and the idea of weekly—let alone daily—deliveries sends shivers up the spines of many manufacturing veterans.

The fact is, you can adopt MRP II and JIT/TQC without tearing your company apart, by making gradual changes to the way you presently do business. Today's best manufacturing companies all operate with MRP II and JIT/TQC, but they didn't simply incorporate the tools overnight. For instance, when Xerox reduced its supplier base from more than three thousand to less than three hundred, it did so gradually, over a period of several years. No successful company unplugs the old tools on Friday and powers up the new ones on Monday morning.

Consider starting to work on MRP II immediately; if your company isn't currently operating under MRP II, spearhead an effort to get it in place—you'll see the positive effects on purchasing in less than a year. You can also begin some JIT activities once your MRP II efforts are under way. Perhaps you can consolidate your supplier base, spreading your volume over fewer suppliers. Then start more frequent deliveries, maybe shifting from once a month to once a week. Once you've gotten accustomed to this way of doing business, you can begin working with your best suppliers to improve quality so that eventually you're able to eliminate inspection of incoming materials. Continue gradually until you accomplish your full range of goals and achieve purchasing excellence at your company.

As you implement the new tools of the trade, you'll be taking a major step toward reducing the four dollars your company is spending on purchased materials for each dollar it spends on direct labor. And in today's highly competitive manufacturing arena, every cent counts.

SUMMARY

- In traditional settings, schedules are not valid and dates are constantly changing. Buyers are typically not willing to be held accountable for on-time deliveries against the dates they get on requisitions.

- Because of competitive pressures, companies are spending an increasing amount of money on direct purchases relative to labor. The financial incentives for reducing the cost of purchased materials are significant.

- In traditional settings, communication "merry-go-rounds" waste valuable time and energy.

- In a traditional environment, buyers are put in a double bind: "Save money, but don't short us on materials."

- In an MRP II/JIT/TQC environment, most problems that exist in a traditional purchasing environment simply disappear as communication improves dramatically, valid schedules replace best guesses, waste is eliminated, and the company seeks a course of continual improvement.

Chapter Two

Profile of a Good Supplier

UNITED YOU STAND

As more and more companies are coming to realize the importance of win-win relationships with their suppliers, it's become popular to think of good suppliers as "business partners." For companies to succeed with Just-in-Time and Total Quality Control programs, this kind of thinking is essential, because suppliers must consistently provide the highest possible level of service.

In legal terms, a partnership is a voluntary agreement between a number of individuals who decide to conduct business together. Partners are required to supply one another with accurate and complete information about anything that affects the working arrangement. A partnership also entails a commitment to use available resources to achieve the common objectives, as well as a joint sharing of risks and liabilities. The commitment to using these resources and the willingness to share in the risks means that the partnership entity will be an ongoing relationship that will endure bad times and thrive in good times.

The same holds for partnerships between companies and their suppliers. The relationship must be open and honest in terms of information; it must be based on a commitment to use available resources to achieve common objectives; and it must share the risks as well as the rewards. Finally, a partnership between company and supplier is a long-term proposition, intended to endure the inevitable highs and lows that all businesses encounter.

Let's say a supplier is willing to enter into a partnership arrangement, as described above. This partnership is a two-way street. Therefore, the buyer and the supplier must change the way they do business if the relationship is to be win-win and long-term. Chapter 10 covers the qualities a buyer needs to possess in order to be a suitable member of the team. The present chapter covers the qualities the supplier needs. Based on the collective experience of many thousands of companies, we can paint an ideal supplier profile based on nine key criteria:

1. Delivery
2. Quality and reliability
3. Price
4. Responsiveness
5. Lead time
6. Location
7. Technical capabilities
8. R&D investment plans
9. Financial and business stability

Each of these points is discussed below. We have developed a separate list of questions for evaluating your suppliers to help determine if they are capable of delivering the level of service and performance you should be looking for in each of the above nine areas. Those questions appear in Appendix A.

1. DELIVERY

Just in Time Means on Time

In a JIT environment, material must be on hand when it is needed. There are two reasons for this. First, the inventory on all items is dramatically reduced, and if the supplier were to deliver late, the company would run the risk of running out of the material. Second, the items are often delivered by the supplier directly to the location where they are to be used. Typically, there is very little storage area for the parts in the production area, and if the items were delivered too early, say two weeks before they were needed, there would not be room to store them. Therefore, it is not unusual for JIT companies to require their suppliers to deliver product within a very tight window; the supplier can deliver a maximum of two days early, but never late.

This is very different from the traditional method (see Figure 2.1) in which material is delivered to the receiving area by the supplier, is then moved to a receiving inspection area where it is checked by quality control, and then moved to a store or warehouse location. At a later date it is then moved to the manufacturing floor, or to a remote location where reconditioning or remanufacturing operations are performed, or to a distribution center where it may be sold as a spare part.

None of these traditional steps add value to the product—they just add cost. By suppliers delivering material directly to where it is used in the manufacturing process (see Figure 2.2), you eliminate all of the waste incurred through the traditional method. You can do this because your supplier is certifying the quality and quantity of the delivered goods. A good supplier is one who delivers material just when it is needed, in the quantity needed, directly to the location where it is needed on the manufacturing floor, or to remote distribution centers, and certifies the quality.

But what happens if the supplier can't make a particular delivery date? Ours is not an ideal world and unforeseen problems are bound to occur. In the event of a problem, a good supplier will be willing to notify the buyer as early as possible of the difficulty. Bad news is bad news— the sooner the buyer is informed, the sooner the company can develop alternative plans and reschedule accordingly.

Finally, delivery performance generally improves when it is monitored. Someone once said, "Anything that gets measured gets better." If we don't measure where we are and establish where we would like to get (98 percent on-time), it is very difficult to say whether performance is acceptable or not. (Chapter 12 covers how to measure supplier performance.) A good supplier will monitor his own delivery performance and work with his customers to see how they can jointly remove the obstacles that prevent 100 percent on-time performance.

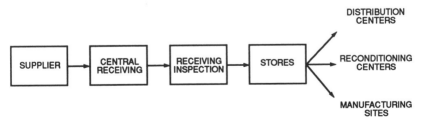

Figure 2.1

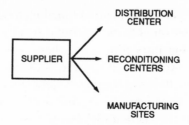

Figure 2.2

Short and Frequent Deliveries, Stable Lead Times

If you have Just-in-Time working at your company, suppliers must be willing and able to deliver in small quantities on a daily or weekly basis as specified on your schedules. This is a big change from the traditional Economic Order Quantity concept that recommends buying in larger quantities to get a price break, but it is essential to making JIT work. And as we will explain later in this book, smaller quantities don't have to translate into higher prices—they should be offset by the better planning information you're offering the supplier and the opportunity to become more efficient at manufacturing your items through methods such as repetitive flow processing.

To achieve the kind of delivery schedule you need, the supplier must be willing to commit to short lead times. As we will explain later in this book, by "short" we mean "days," maybe two or three weeks at the most. Equally important, the lead times must be stable. Since you're providing the supplier with your schedule, he should plan his capacity and raw material around your dates. This should stabilize his lead times into the future.

Delivery: Have It Your Way

A good supplier will not only deliver items according to your need date, but will package the items according to your specification; if you want it packed in lots of a hundred, for example, that's how it should arrive. You might also provide the supplier with standardized baskets or tubs.

When you provide the supplier with standardized baskets and instructions on how to fill them with parts, you simplify your receiving process. For instance, if your instructions say to fill the basket to a certain mark on the side that indicates that the basket contains 95 items, then you'll know exactly what you're getting. Your receiving personnel verifies that the 10 baskets they received are all filled to the mark, so they know for certain that they have received 950 parts (10 baskets times 95 units per basket) and do not have to count the parts.

A related issue is how the items are delivered. Ideally, good suppliers (if local) should be agreeable to using their own truck if they have one. If not, they should be willing to use a vehicle or carrier that you designate. In addition, whether it's their own truck or a commercial carrier, good suppliers should make sure that the same driver makes the run each time, and that the delivery will be made to the point-of-use as specified above. The reason for using the same driver is that you will eliminate any confusion about where the material should be delivered, and avoid possible delays caused by items being dropped off at the wrong location.

Flexibility

The marketplace can change very quickly, and the best-laid plans and schedules may have to be changed at the last minute to seize opportunities or avert crises. You and your supplier need to understand and agree that change is the only constant in the world of business, and that flexibility is the key to survival. Good suppliers would be willing to review their processes and try to build in the flexibility you need.

A good example of this flexibility is the area of sand castings. Current lead times on sand castings run ten to fourteen weeks, but in actual practice, the supplier packs the sand around the core and pours the metal into the mold the week they ship the product to the customer. Let's say that your company purchases four similar configurations and uses a total of one hundred castings per week. Your company would like the supplier to reserve capacity for one hundred molds per week (per the supplier schedule) and let the buyer choose one week before the delivery is required which configuration is to be put into the mold. This same logic holds if your company is buying plastic molded parts, chemicals, etc. If the supplier is able to build in this flexibility, he can more easily respond to customer and marketplace changes.

2. QUALITY AND RELIABILITY

Toward Zero Defects

Correct quality means that the good supplier ships materials according to your agreed-upon specification. The good supplier is in conformance to your requirements not only on the first delivery, but on each and every delivery for as long as the partnership lasts. That conformance includes not only the physical requirements of the specifications, but all the other requirements as well. Correct quality would also include correct quantity. When you order 100, the good supplier would ship 100—not 95 or 105. If the shipment is the exact amount you requested and the supplier's paperwork shows the correct quantity, you should not have to count the items when they arrive at your receiving dock. If the quality is correct every time, you should not have to inspect the items. If you don't have to count or inspect the items, the good supplier can deliver the items directly to the point-of-use on the manufacturing floor.

The buyer will want to certify that the supplier is capable of delivering the correct quality. The first step is to sit down with a representative from the supplier and discuss the specification. Next, the buyer will want to bring his or her quality-assurance people into the supplier's plant, walk them through the supplier's manufacturing process, and verify that the supplier can make the material to your specifications. Once the buyer's quality-assurance people verify that the supplier has the capability to produce the desired quality level (i.e., the equipment and the understanding), they will probably need to check only the first few shipments. If the materials meet the buyer's specifications, from that point on the supplier is accountable for shipping only good parts and the buyer should not have to inspect the parts when they arrive at the receiving docks. Good suppliers will welcome the buyer's audit, will want to be certified on quality, and will conform to the requirements of the buyer. Chapter 6 discusses supplier quality assurance.

Feedback

To help the supplier stay on target, the buyer should request that the supplier send the quality measurements for each lot the supplier produces. The measurements, which should be sent with each shipment, will reveal whether the supplier's quality is stable or a negative trend is forming. These measurements, all the key specification requirements requested by the buyer, are usually plotted on a control chart.

Figure 2.3 shows a control chart. The particular requirement being charted has a range of .88 to .92, and each "X" represents a test result of a lot of material. In Figure 2.3, the control chart indicates that the supplier is in danger of going out of range. If the supplier does not correct the cause that is allowing the particular test requirement to deteriorate, the next lot could be .93 and out of range. Maybe the tool isn't sharp anymore, or the process isn't being run correctly by the

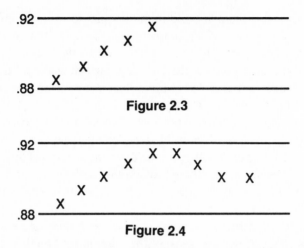

Figure 2.3

Figure 2.4

operator—it could be any one of a hundred different reasons. Figure 2.4 indicates that the supplier has corrected the problem and, likewise, the trend has been corrected. A good supplier is willing to chart the process to correct deviations in order to consistently deliver product that is acceptable within the tolerances established.

To ensure ongoing high quality, the good supplier should have a monitoring and measuring system in place. It is not unusual for companies to occasionally visit their suppliers and perform random spot checks to ensure that the supplier is following established procedures and, in fact, monitoring the process. A good supplier would welcome these random checks and not interpret them as a lack of trust, but rather as an essential element in a cooperative team effort to ensure good quality. A good supplier would also perform random checks of their own facility to assure that their quality-assurance process is indeed producing the desired results.

People will come to appreciate the value of feedback and regular communication between your quality-assurance people and the supplier's. When this two-way communication becomes a way of life, quality becomes a way of life. Highly visible quality programs promote this kind of thinking, and are therefore an important criterion for selecting a good supplier.

Building Quality into the Product

Another important aspect of the good supplier's program is "controlled design." Good suppliers should have documented policies and procedures that enable them to build in quality, not to wait and inspect the items after a production run is complete. For example, if the supplier has a run of a thousand parts, then inspects them *after* the last one is finished and finds that the first is bad, the odds are that the whole lot is bad and they will have to start over. In other words, the entire run is wasted, and everyone has lost time and money.

In contrast, if they have a process control in place, they'll verify quality as they go along. If a problem is detected and cannot be immediately corrected, they'll authorize the operator to turn off the machine. It's therefore important for the good supplier to have a policy in place that enables an operator to shut down the equipment and correct any problems that affect quality.

It is sometimes difficult for managers in manufacturing companies to accept this concept, because most of them are judged on their ability to get product out the door. Consequently, the idea of giving people line authority to turn off a machine might be tough to swallow. But the alternative—taking good raw material, good labor, and making bad parts—makes no sense either.

Resistance to this concept dissolves when people understand that quality is something that must be built into the system, not checked for afterward. That starts in the beginning, with incoming material. If suppliers start off with substandard or defective material, they can only make substandard or defective products. The old computer acronym, GIGO—"garbage in, garbage out"—applies equally well to a manufacturing company. Good suppliers will have documented programs for building quality into the parts they produce.

Maintaining Top-Notch Service

Of course, ours is not an ideal world, and problems will occur even in the best environments. Good suppliers will recognize the inevitability of problems and will have the equipment and capabilities in place so they can quickly recover from problems as they arise. Remember, in a JIT environment it is unacceptable to have defective purchase items or delays. JIT operates under the same principle as a series of dominoes: If your supplier is late, it could shut you down; if your supplier gives you defective material, it could also shut you down, and you want to minimize the chance that your supplier(s) will initiate a negative chain reaction.

One way to minimize the likelihood of a negative chain reaction is to make sure that suppliers understand that surprises are not acceptable. That is, they cannot change any raw materials or their manufacturing process without telling you first, because the effects could be devastating.

This is well illustrated by the case of a manufacturer that had been purchasing the same ingredient used in processing vinyl for many years. All of a sudden, the vinyl manufacturing process wouldn't generate acceptable product. The reason? The supplier of the ingredient didn't tell the manufacturer that he was using a new raw material. And the new material, it turned out, contained a trace element that caused the end process to go out of control. This "sin of omission" on the supplier's part cost the manufacturer a great deal of time and money (and almost caused the supplier the business). Had the supplier simply told the vinyl manufacturer of the change, the vinyl maker could have run a sample and quickly determined that the new compound wouldn't work.

A Teamwork Approach

A good supplier understands that quality programs and troubleshooting require teamwork, both within his company and between his company and yours. It doesn't matter who's at fault; the focus must be on solving and eliminating the quality problem. The supplier must therefore adopt the attitude that there's no such thing as problems—only opportunities for improvement. He must firmly believe that as his people work together with your people as a team, they'll learn from mistakes and improve. Moreover, they'll be able to provide better service to their other customers, too, so the benefits will continue beyond their dealings with your company.

Full-Service Capabilities

You will want to find suppliers who have the ability to take an idea and then design, as well as manufacture, the necessary items. Some suppliers are excellent at designing quality items. They incorporate quality planning as a regular part of the design function, and design all items with an eye toward manufacturability. Unfortunately, their manufacturing processes may not be capable of producing quality parts at the level you desire.

The opposite situation may also exist. The supplier may not have the ability to design items for you or make suggestions as to how to design a part so it is more manufacturable. But given the proper tooling by the buyer, that supplier may be able to consistently produce good parts for you.

The most desirable supplier will be full-service, one who can take your idea and design a quality part that can be manufactured to the desired quality level. The good supplier should then have the ability to build the required tooling and test equipment to assure the quality of the part when it is produced. Finally, a good supplier should be able to use the tooling and manufacture defect-free parts that consistently conform to your specifications, comply with your delivery schedules, and at the same time maintain minimum costs. Because a full-service supplier has all of the above capabilities, he can truly be an asset to your company as you develop new products and improve your current offering.

3. PRICE

Fairness for All

Traditionally, the buyer's job is to hammer the price down as low as possible. The supplier then accepts the lowest possible margin out of fear of losing the business, and tries to figure out how to cut corners and regain their profit. With Just-in-Time, such vicious tactics are eliminated in favor of a "fair" price. It may not be as low as you would like or as high as the supplier would like. But a fair price is one that works for both parties. "Fair" means that the supplier can make a profit, but you can still remain competitive in the marketplace. In other words, a fair price represents a "win-win," mutually beneficial situation.

To arrive at a fair price, a good supplier will reveal how he establishes profit margins. If you understand the elements of the supplier's costs and pricing, you can actually help the supplier reduce costs by having your engineers assist in making his processes more effective. The reverse of this is also true: The supplier's engineers can assist you in the design of the item to lower the cost.

An office furniture manufacturer used a self-leveling glide on the bottom of certain desk models. The glide, purchased from an outside supplier, used to consist of nine separate parts. To reduce the complexity of the glide, the manufacturer sent a group of its engineers to the supplier's company to redesign the item. The engineers, working hand-in-hand with the supplier's engineers, found a way to cut out four parts. This reduced the unit cost of the glides from twenty-five cents to twelve cents. The supplier passed the lower cost on to the manufacturer while still retaining his desired profit margins, and further profited by applying what was learned in the process to other customers as well.

This example shows how you can perform value analysis that asks, "What makes it work?" (the primary function) and "What makes it expensive?" (the secondary function). If we can the eliminate the secondary function while maintaining the primary function, costs will be effectively reduced. Another way to look at it is, "How can we define the part better so it does what it's supposed to, but at a lower cost?"

In for the Long Haul

In addition to a willingness to accept a fair price, a good supplier will agree with the buyer to a long-term commitment, which means at least two, three, or four years. If you can tie the pricing over a long-term agreement, both of you will benefit. This should help ensure that you will get the same price over time regardless of the individual release quantities, because the supplier can plan raw material and capacity well into the future and take advantage of economical purchasing and manufacturing opportunities.

Another aspect of long-term commitment is the agreement to future price reductions. Over the term of the contract, you want to work together with your supplier to reduce his costs so that at the end of the contract the purchased parts will be less expensive. At first blush, this defies common sense. After all, over the course of time, the supplier will experience increases in raw material or labor due to situations that may be beyond their control.

Increases in the cost of raw materials, though, don't have to translate into higher prices for finished goods. First, you want to work with a supplier who will inform you of price increases in raw materials as they occur. You can then work together to tackle the problem. Perhaps your people will suggest an alternative process that will still yield acceptable parts. Or, if the supplier's increase is substantial, you might authorize him to make a substantial purchase of raw materials in advance to hold off the increase. While this may add to the supplier's cost in the short run, it may allow you to accept a very small increase in the price now and delay a substantial increase later. If the supplier gives you advance notice of price increases, you have time to determine if alternate actions would be appropriate.

One company was notified by a paint supplier that a substantial price increase was about to go into effect due to an increase in the solvent cost. The supplier was willing to commit to more inventory but didn't have any place to store the materials. So the buyer paid the demurrage charges to let the solvents sit on a rail car, since the cost of doing so was actually less than the price increase. Now, that's a creative partnership arrangement!

4. RESPONSIVENESS

Who's in Charge Here?

Is there someone here in the supplier's Order Entry or Planning Department who's assigned to my account? Does that person know my items? Is there someone in Customer Service who can follow through with problems and complaints and get back to me very quickly? Is there a technical service representative assigned to my account? Where can I call that person if I have a problem? Who is the quality contact—can I call that person by name if I have a problem?

These are the kinds of questions that you'll want answered before doing business with a supplier. A good supplier will never dismiss the issue with "Call Customer Service if you have a problem." Instead, he'll provide an organizational chart with telephone numbers.

The supplier should also inform you of whether there are field and technical support people in your area who can quickly respond if you have a problem. It's vitally important that the supplier gives you an honest assessment of the skill level, strengths, and weaknesses of the individuals who might be servicing your account. The reason you want to know the skill level is that if you have a technical problem in an area in which the field support person is weak, that person may give you an incorrect answer or may have to call someone else to get the solution. You want to be able to talk directly to someone who will understand your problem and be able to give you a good solution. In short, when you call for help, you don't want to get the runaround. The attitude again must be, "Let's work together as a team."

5. LEAD TIME

Since lead times are one of the keys to competitiveness, your suppliers must be willing to discuss their true lead time situation: How long does it take them to get raw material? When do they set aside capacity? How long does it take them to actually make the item? In other words, what are the elements of the supplier's lead time?

A good supplier will explain how he schedules his key suppliers. This is important, because if one of your customers makes a special request, you must know whether or not your supplier can comply with it. If the supplier stocks the necessary items, you know you can make the changes quickly. But if the supplier requires three months to produce the necessary material, your special needs probably won't be met. In that case, you need to understand your supplier's—and hence your own—constraints.

Take fabric manufacturing as an example. The quoted lead time today is ten weeks, which is made up of four weeks to order the yarn, four weeks to weave the "greige goods" (the undyed fabric), and two weeks for dyeing. If the supplier is willing to allow you to make color changes at the two-week window, it eliminates a constraint in responding to last-minute customer needs. In this example, the buyer buys the capacity to weave the fabric needed, authorizes the supplier to buy the necessary yarn, and specifies the color at the two-week window.

Another issue, and a key one in reducing and managing lead times, is the supplier's willingness to work with you as a partner. Are there technologies that you employ in your plant that the supplier doesn't (i.e., quick tool changeovers, cellular manufacturing, etc.)? Is the supplier willing to let your employees visit his facilities and assist in reducing lead times?

One small manufacturer of molded Plexiglas sheets employed no engineers and, as a result, had a very inefficient plant layout and some very long tooling changeover times. Because of this, the supplier's manufacturing lead times on the sheets were ten days. The buyer's company sent a group of engineers, tooling designers, and quality personnel to the supplier for a two-day visit. They suggested how the supplier could change the layout of the equipment on the plant floor, and how tooling changeover times could be reduced. As a result, the supplier was able to reduce the manufacturing time to *one* day!

How receptive are your suppliers to working on lead-time reduction? Good suppliers should be eager to reduce their lead time in any way they can.

6. LOCATION

An Open-Door Policy

If a supplier has multiple manufacturing locations and is honest regarding his capabilities, he'll reveal the capacity and other vital numbers of each location. Are all of the plants equally efficient? Do they all produce the same quality level? If not, do you have the ability to choose which factory will be used to make your material?

If the supplier subcontracts, who are key subcontractors? Where are they located? Will the supplier help you talk with them so you can determine their strengths and weaknesses? This is important so you can be assured that quality will be maintained throughout the supplier's operation.

Part of the supplier's open-door policy involves revealing all of the other products he manufactures for other companies. How is the supplier allocating resources to your product versus other products? You must be able to know that at a particular work center, your product will be a certain percentage of the workload. Let's say a supplier is 60 percent sold out, except for one drill press, which is used for your product, and that drill press happens to be 100 percent sold out. Another customer places an unusual order, which requires your drill press. If that order exceeds the capacity of your drill press, your order may suffer as a result. A good supplier will honestly explain how he allocates capacity among his customers. That information is essential to knowing what kind of service you can expect over the long haul.

Finally, what is the overall quality level of the product that the supplier offers to his other customers? If he is an excellent supplier, you can make an assumption that he could make your products just as well.

7. TECHNICAL CAPABILITIES

In Search of the Cutting Edge

When it comes to technology, suppliers come in two varieties: leaders and followers. Obviously, you want a supplier who's on the cutting edge of technology and is seeking better ways to manufacture their products. This will allow you to continuously improve your products in terms of quality and performance. If your supplier is a follower, he'll let someone else do the ground-breaking. Sure, the followers might offer slightly lower costs because they don't have to amortize the R&D and capital investments, but their products will never be cutting edge. Which means your products may never become cutting edge either.

Besides being on the cutting edge of technology, good suppliers have good track records for bringing in new products on time. This is important, because if they can introduce their own new products and new products for other customers on time, they can introduce your new purchase components on time, too.

A related issue is how well suppliers have documented their processes. Are the supplier's successes hit-or-miss? Or are they part of a well-thought-out development strategy for bringing out new products and product improvements each year? If suppliers have such a strategy, chances are they have the resources to improve and enhance the products they make for you, too.

The same goes for product improvements. Do the suppliers' plans call for continual improvements in their product lines? Have they allocated money to improve your products? Are they continually upgrading and adding equipment? Do they have a sufficient number of people to support the improvement efforts? The supplier who can answer "yes" to all these questions is the kind of supplier you want on your team.

What about the supplier's history of success with new products and product improvements? The supplier should be willing to show you reliability data, problems and all. You also want to get a satisfactory answer about how problems have been handled in the past. What kind of process have they used to find solutions? Who did they assign to troubleshoot and correct the problems? These are important questions that a good supplier will readily answer. Finally, you want to see that TQC is a standard operating principle for the supplier's R&D team. If total quality is considered as early as the design phase, it will likely be built into the entire production process.

8. R&D Investment Plans

Funding Tomorrow

Good suppliers will have an R&D budget to improve old products and develop new ones. If they haven't set aside funds, then they aren't going to carry out any R&D activities. But R&D can also be a double-edged sword; after all, if a supplier is so wrapped up in his own or someone else's R&D efforts, will he have time for yours? You want a supplier who will be able to answer that question honestly.

Another important aspect of the R&D issue is whether the supplier is willing to participate in joint design programs. A good supplier will be willing to work with you in exploiting the creativity, technical expertise, and production knowledge that both firms bring to the party.

Finally, the kind of suppliers you want on your team will be responsive to your development needs. When you suggest a new project or joint effort, they'll be eager to sit down and discuss the details.

9. FINANCIAL AND BUSINESS STABILITY

Here Today . . . Here Tomorrow

Clearly, you only want to trust your business to a company that has the stability required to survive tough times. Why spend six months qualifying a supplier only to find out he's in Chapter 11? This seems obvious, but it is too often overlooked. Even if a supplier doesn't slip into the abyss, if he is teetering on the edge, he won't be able to provide the kind of quality and development services you need to be a world-class competitor.

In addition to checking the basics—Dun & Bradstreet rating, industrial average data, profitability for the last three years, return on equity, debt to equity, liquidity and solvency, and receivables turnover—you'll want to know whether the company is running on borrowed money or revenues from sales. You would also like to know if there is a high turnover in the management team or an expected change in ownership that could affect your long-term relationship. A good supplier will provide all the necessary information you need to assess his financial and business strengths and weaknesses.

SUMMARY

- A good supplier will build quality into the product, aiming toward zero defect production. This requires a commitment to quality, and a teamwork approach that entails feedback and the spirit of cooperation.

- Delivery performance is a key measure of a good supplier. Desirable delivery performance not only entails hitting a schedule, but involves a willingness to make short and frequent deliveries to point-of-use areas within your facility, and a willingness to package items according to your specifications.

- Good pricing means fair pricing; ultimately, it may not be as low as you would like or as high as the supplier would like. But everyone wins with a fair price because it buys you top-notch and sustained service, and assures the supplier ongoing business.

- A good supplier will demonstrate responsiveness to your needs by ensuring that readily accessible people are in charge of servicing your account.

- Long lead times are the enemy of all businesses; a good supplier will work with you to reduce lead times as much as possible.

- It is vitally important to know a supplier's capabilities and work load at various locations. A good supplier willingly provides you with all necessary information regarding their locations.

- The best suppliers create the future rather than suffer through it. Look for suppliers on the cutting edge of technology.

- A cutting-edge company will reinvest some of its profits in R&D; you want a supplier who has a long-term view and is willing to spend for tomorrow.

- All suppliers must meet the stringent financial stability criteria you would use to evaluate potential new customers for credit.

(See Appendix A for a list of questions that will help you identify good suppliers.)

Chapter Three

The Basics of Supplier Scheduling

One of the keys to effective purchasing is to have a valid schedule; that is, one that specifies what and when materials are needed. The typical buyer is swamped with past-due orders that no one needs, and at the same time is being asked to expedite in parts with a due date of next week. When asked whether he or she would like to be held accountable against the dates on the purchase orders and hired or fired based on 95 percent on-time to the dates on the requisitions, any sane buyer would answer, "No way!" Why? Because the dates aren't valid need dates; they are typically replenishment dates, two or three weeks before the parts are needed in production.

In this setting of invalid schedules, buyers are now being asked to bring in the material Just-in-Time. Unfortunately, it doesn't make sense to ask the supplier to ship more frequently, say weekly, to an invalid schedule. Why bring in parts more frequently just to see them sit on the shelf for several weeks before they're needed?

Despite the lack of valid schedules, buyers are also asked to participate in total quality-management programs that move toward zero defects on all supplier items. The program is put in place and all the supplier education is completed. Quality is emphasized over and over. But because of invalid schedules, the buyer is asked to expedite the material in two weeks earlier than requested because it's needed this week. The supplier rushes the material through the system and is able to deliver it in two days, at which point it is rejected for poor quality. Since the materials are necessary for production, though, the rejection is waived and the defective materials are used. This, of course, gives a double message about quality: Formally, the company demands zero defects, but in reality it will accept something far less. Thus, without valid schedules, it is impossible to do a good job of Just-in-Time or Total Quality Management.

BUSINESS AS USUAL REVISITED

We previously described the flaws of the traditional communication process between supplier and buyer. In this chapter, we'll revisit the traditional communication pattern, then describe a more effective means of carrying out purchasing: supplier scheduling. To recap the traditional flow of information, it begins when the planner requisitions the buyer. The buyer then sends a purchase order to the supplier's Sales Department, and the Sales Department communicates with the planner in the supplier's plant. This is summarized below.

If a problem arises at the supplier's end and he can't comply, the planner at the supplier's operation notifies the salesman, who notifies the buyer at the customer company, who then notifies the planner. This reverse flow of information looks like this:

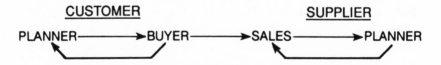

Now let's say that the revised promise dates aren't satisfactory and will seriously affect the production schedule. The planner so advises the buyer, and so on as shown below:

This communication flow is not only inefficient, but it's unnecessary. It's also the method most frequently used in American industry today. Why? Because it grew up that way. Back in the days of manual inventory systems, only the people posting the inventory records knew the requirements, and it didn't make sense to have the buyers also posting inventory records. The inventory clerks (planners) therefore sent requisitions to the buyers, who did the ordering.

THE SUPPLIER SCHEDULING CONCEPT

Letting Buyers Be Buyers

With computers and formal systems that work, it's no longer necessary to do purchasing the old way. Instead, it makes more sense to put the MRP planning people in direct contact with the supplier. These people are called "supplier schedulers," and they typically communicate directly with scheduling people at the supplier's plant. It is their job to take the valid schedules generated by MRP and work with suppliers to be sure those schedules are attainable. (Chapter 5 explains their roles in greater detail.)

With supplier scheduling, the buyers are freed from the ongoing pressure of expediting and paperwork, and as a result have the time to do the really important parts of their jobs: sourcing, negotiation, contracting, value analysis, etc. See Figure 3.1.

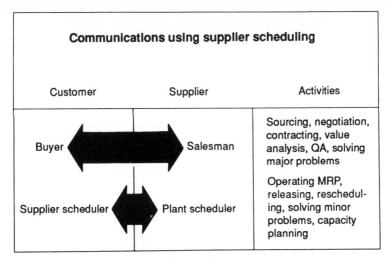

Figure 3.1

As you can see, the buyers buy from the suppliers and the supplier schedulers schedule the suppliers. In this environment, there is normally a business arrangement, called a "supplier agreement," between the company and the supplier. This agreement, as a rule, is the result of efforts by the buyer (for the company) and the salesman (for the supplier) and, possibly, their managers.

The supplier scheduler enters the picture after the supplier agreement has been finalized. The supplier scheduler's job is to operate Material Requirements Planning and provide the suppliers with schedules within the context of the supplier agreement. When near-term requirements change, the supplier scheduler works with the supplier's people to implement the reschedule. When the supplier has a problem meeting the schedule, he notifies the supplier scheduler, who can advise the supplier of the exact requirements and help develop an alternative plan.

Take the following example from a Fortune 500 company, of a phone conversation between Jerry, the supplier scheduler, and Sid, Jerry's contact at the supplier's plant.

SID: Jerry, we've hit a problem running your order for Item #28184.

JERRY: Let's see. . . . That's for a thousand pieces due next Monday, right?

SID: Right. We're less than halfway through the run and the machine went down. It'll take about a week to fix, so I'm afraid we'll be late.

JERRY: Hold on, I'll check MRP II.

Jerry checks the MRP II display (see Figure 3.2) for Part #28184. He notes the following:

1. There are requirements for 250 in Week 2 and 750 one week later.

2. The order ("scheduled receipt") for a thousand due in Week 2 was intended to cover both these requirements.

JERRY: Sid, how many good ones did you get before the problem happened?

SID: 350 for sure, maybe 400.

JERRY: Fine, ship me those. When can you get the rest?

SID: In a week or less.

JERRY: Fine, no problem. By the way, Sid, how's the family?

Part #28184

| | Weeks | | | | | | | |
	1	2	3	4	5	6	7	8
Projected gross requirements			250	750					
Scheduled receipts			1000						
Projected avail. balance	0	0	750	0	0	0	0	0	0
Planned order release									

Figure 3.2

Elapsed time for the conversation: thirty seconds. Contrast that with the traditional flow of information, which probably would have been mapped as follows:

Even in the unlikely event that all of these communication steps are executed successfully, the buyer has spent valuable time playing messenger, time that could have been spent saving the company money.

An End to Endless Paperwork

Questions: With supplier scheduling in place, what happens to the need for written purchase requisitions? Answer: It goes away. Requisitions are the traditional common linkage between two people: the person operating the inventory replenishment system (the planner) and the person in direct contact with the supplier (the buyer). Those two functions are now combined into one person: the supplier scheduler. If the supplier scheduler were to generate a requisition, there would be no one to send it to other than himself. The reason is that with supplier scheduling, purchase requisitions for production material are no longer needed. And as that need for requisitions vanishes, so do hundreds and hundreds of pieces of paperwork.

BENEFITS OF SUPPLIER SCHEDULING

In Chapter 1, we described the most significant benefits of supplier scheduling. The following summarizes the main ones:

• Low inventories. Manufacturers can get just the amount they need, just when they need it.

• Low prices. Buyers have negotiated pricing in advance, across the board. They've asked the supplier for the best price on the entire volume of business and no longer need to be concerned with the details of the traditional pricing structure on a given item for a specific order. Also, price breaks, in the traditional sense, are not as significant as before. Price breaks exist largely to enable the supplier to run a larger quantity at one time, in order to amortize the setup costs over a larger quantity.

The information and visibility contained within the supplier schedule gives the supplier a far better opportunity for cost reduction and efficiency, via techniques such as combining runs, sequencing runs to minimize changeover time, etc. It then becomes the buyer's job to ensure that the supplier passes on a good percentage of these savings. The supplier benefits, and so does the company.

• Constant pricing. Another result, in addition to lower prices, is constant pricing over a longer period of time. Price increases based on "price at time of shipment" are "surprises" that most buyers like to avoid. Since the supplier is authorized to acquire material out into the future with supplier scheduling, price increases based on rises in the supplier's material costs should be sharply reduced in the near term.

If the supplier's labor costs are scheduled to increase, the company may wish to authorize the supplier to run the entire schedule prior to the increase, but to ship the material as needed. Again, this gives the company constant pricing over the life of the schedule.

• Shorter lead times. Most of a supplier's lead time is often made up of backlog. And as the backlog increases, so do the quoted lead times. With supplier scheduling, however, quoted lead times aren't nearly as important. The supplier schedule extends out into the future, well beyond the supplier's quoted lead times. This is a "win-win" situation. The supplier has more and better information with which to plan, and the company is freed from issuing purchase orders for specific items in specific quantities far into the future.

With supplier scheduling, the most important supplier lead time is the finishing lead time. This is never more than the complete manufacturing time, and is almost always less than the "quoted lead time." It represents the point of commitment to specific finished items. This finishing lead time, plus perhaps a week or so for the supplier's internal scheduling, establishes the "firm zone" on the supplier schedule (this is often referred to as the firm time fence).

In the example shown in Figure 3.3, this zone is the first four weeks. All quantities shown within that period are firm. The supplier scheduler resists making changes in this zone, and supplier concurrence is required prior to making a change. Thus the supplier is assured of stability in the schedule, and the company receives the benefits of shorter lead times.

The different items bought from this supplier are listed vertically on the left side of the chart. Time is expressed in weeks, horizontally. The first four weeks are shown individually; the next four weeks are grouped, as are the following eight weeks.

The quantities shown in the columns are the customer's requirements for the various items. Note the asterisks on the requirements in the first four weeks—they represent firm requirements. This four-week firm period reflects the supplier's finishing lead time plus an allowance for

Jones Company supplier schedule for: Smith, Inc.						
Part # Week:	1	2	3	4	5-8	9-16
13579		100*			100	200
24680	20	20*	20*	20*	80	160
42457	300*			200*	800	1600
77543			40*		40	

Figure 3.3

*Firm

scheduling time. In this example, the supplier agreement (not shown) specifies a four-week firm horizon, which means:

1. The supplier is required to ship according to this schedule.

2. The company cannot change the quantity or timing of these requirements without approval from the supplier inside the four-week window.

3. The supplier is authorized to produce any of these firmed-up requirements early and hold them in inventory until it's time to ship. The company is obligated to take them, but not sooner than the schedule specifies without authorization. The supplier is then able to group production runs to achieve internal efficiency. The company receives shipments in quantities that directly match the production requirements and, if the buyer chooses, on a "Just-in-Time" basis.

To take this example further, the supplier agreement might also specify that requirements in the second four-week period represent a commitment to cover the supplier for any material unique to that customer. In other words, the supplier is authorized to purchase special material for these requirements. If the company's requirements are eliminated for all items using a special or unique material, the company may have agreed to protect the supplier from being saddled with obsolete stock. This could be done, if necessary, by paying the supplier's restocking charge from his supplier or, worst case, directly reimbursing the supplier for the costs of the material.

The column furthest to the right, representing Weeks 9 to 16, contains information only, and does not reflect any company commitment. The information column is mainly to help the supplier plan resources— people, equipment, tooling, money, etc.

As we mentioned in the previous chapter, this is one of the characteristics we look for in a good supplier, the willingness to work to shorter, stable lead times.

• Reduced paperwork. For the reasons described above, supplier scheduling greatly reduces the paperwork burden.

THE SUPPLIER SCHEDULE: CONTENT AND FORMAT

A purchase order is simply a contract with a schedule on it. This doesn't make a lot of sense, because schedules change frequently, every week or so, while contracts change infrequently, perhaps once a year. There's little reason to reissue a new contract every time the schedule changes. So why not separate the schedule from the contract? An increasing number of companies are, in fact, doing just that. They establish long-term supplier agreements and then issue schedules, typically once per week. In general, the buyer develops the supplier agreements, while the supplier scheduler takes care of the schedules. See Figure 3.4.

The dates and quantities shown on the supplier schedule are need dates. These dates are generated by the buyer's MRP system and are based on when the company needs the material to meet its production schedule. The quantities shown with an asterisk are scheduled receipts—they represent open orders. The supplier is authorized to produce and deliver those quantities in the appropriate weeks. The quantities to the right are planned orders; they're recommendations from the MRP system as to when the buyer will need additional quantities, and are provided to the supplier to give future visibility so the supplier can efficiently plan raw materials and capacities. An explanation of how those dates are generated is in Appendix D.

A more detailed example of a supplier schedule is shown in Figure 3.5.

The schedule shows the following:

• The company's name, the supplier's name, the supplier number, the date, the initials of the supplier scheduler, and the buyer.

• The first six columns show weekly data, the seventh shows a four-week accumulation, and the eighth shows a twelve-week accumulation.

	Supplier agreement	Supplier schedule
What	Prices, terms, conditions, commitment periods, etc.	Which ones, how many, and when
Who	Buyer and salesman (and/or their managers)	Supplier scheduler and designated person at the supplier's plant (ideally its master scheduler)
How often	Normally every year or two. Possibly open-ended.	About once per week

Figure 3.4

Jones Company supplier schedule for: Smith, Inc. – week of 02/01/9X

Supplier #114

Supplier Scheduler: AB
Buyer: CD

Firm Zone: first 04 weeks

Material Zone: next 06 weeks

-------- Requirements --------
FFFFFFFFFFFFFFFFFFFFFFFFFFF MMMMMMMMMMMMMMMMMM

Item #	Description		Week 2/1 & previous	Week 2/8	Week 2/15	Week 2/22	Week 3/01	Week 3/08	Next 04 Wks	Next 12 Wks
13579	Plate	Qty: PO#:		100 B1146				100	100	300
24680	Panel	Qty: PO#:	20 B1122	20 B1146	20 B1180	20 B1203	20	20	80	340
42457	Tube	Qty: PO#:	300 B1122			200 B1203			1100	3500
77543	Frame	Qty: PO#:			40 B1180		40			

Figure 3.5

Thus, the supplier in this case is provided with twenty-two weeks of forward visibility. If necessary, more could be provided by merely increasing the number of weeks in the columns farthest to the right. The key is giving visibility to the supplier well beyond his quoted lead times. A recent survey of companies using supplier scheduling indicated that 85 percent of the companies gave their suppliers twelve or thirteen weeks' visibility, 10 percent gave their suppliers twenty-two to twenty-six weeks' visibility, and 5 percent gave either a year or some other number of weeks visibility.

• The firm time fence is set at four weeks into the future. This is indicated in several ways:

1. It's spelled out in the header information.

2. The letter "F," indicating "firm," is shown above the dates for the first four weeks.

3. All required quantities in the first four weeks carry a purchase order number below them. In this example, the company has elected to retain P.O. numbers for administrative reasons, even though it has eliminated individual purchase order paperwork.

The issue of whether or not purchase orders are necessary can be an emotional one in some companies. While individual hard-copy purchase orders are not necessary, P.O. numbers may be needed for control purposes.

One method to accomplish this was demonstrated above in Figure 3.5. Another would be to use the item number and due date. Thus, the first item would have a purchase order number of 13579-0208. A third alternative entails using the supplier number and due date. This would show all quantities in a given week under one number, for example 114-0208.

Companies not requiring purchase order numbers would probably print an asterisk next to the requirement to indicate that the order is firm, as shown earlier in Figure 3.3.

The Degree of Future Commitment

Questions sometimes arise regarding the degree of commitment to the suppliers involved in supplier scheduling. Does a company find itself making firm commitments further into the future with supplier scheduling than with conventional purchase orders? Or further than with blanket orders? Or further than with "buying capacity"?

In each of these cases, the answer is "no." Supplier scheduling, done properly, results in a reduced firm future commitment because it gives the supplier the future visibility necessary to plan the necessary resources to support the schedule. Here are the alternatives:

• Conventional purchase orders require commitments to specific items, quantities, and timing out through the supplier's total lead time. Of course, this includes the supplier's "backlog" time. Supplier scheduling keys on the supplier's true manufacturing time and firms up only those orders within that period. Since these times are shorter, the commitment period is correspondingly smaller.

• Blanket orders typically require a commitment to the supplier for a specific quantity (or perhaps quantity range) of a specific item over an extended period of time into the future—often at a predetermined delivery schedule. In supplier scheduling, there is no long-term commitment to specific quantities and timing.

• Buying capacity usually means committing to the supplier for a given level of volume per time period over an extended period of time into the future. With supplier scheduling, there is normally no such firm commitment. There are situations, however, when good supplier scheduling requires capacity buying, and we'll discuss these later in the book.

Why should the supplier see this as a win-win situation? At your company, would you rather forecast what all of your customers are going to buy or would you like it if all of them gave you a thirteen-week printout of what they needed? Of course you would like their actual schedules. Why are your suppliers any different? The advantage to the suppliers is that you are giving them valid schedules into the future of what you intend to produce week-by-week. Based upon that visibility, the suppliers are now able to better plan raw materials and capacity to hit your schedules. The suppliers no longer have to forecast what your actual week-by-week requirements are, because they can see them on the supplier schedule.

Multiple-Sourcing

Throughout this book, we mention that the trend today is toward fewer suppliers and single-sourcing of components. (See Chapter 7.) There are still situations in which multiple-sourcing is important, however. For example, your volume requirements might be so great that you can't get all you need from one supplier. Similarly, there may be a geographical issue: You might have an East Coast and West Coast operation, both using local suppliers—why ship product three thousand miles when the parts you need are three miles, thirty miles, or three hundred miles away?

Another reason for multiple-sourcing is the government. For example, the FDA requires that pharmaceutical companies making new drugs identify all of the sources for the active ingredients. Moreover, those sources—even backups—must make a certain number of batches per year, which means that for safety's sake, you'll have to be dealing with multiple sources to satisfy the requirements. In some cases, the Department of Defense also requires multiple-sourcing to meet its regulations.

In an ideal world, you might be able to single-source. And JIT will always force you toward that goal. But the reality is that for one reason or another, most companies have to multiple-source in one way or another. Fortunately, supplier scheduling does not require single-sourcing, and most of the companies using it today are multiple-sourced in significant ways.

Good software for supplier scheduling makes multiple-sourcing more practical. With good software, the supplier scheduler is able to specify the percentage of an item's total requirements for Supplier A, Supplier B, etc., and the supplier schedules reflect this split. (The supplier scheduler is typically not a decision maker in this matter of apportioning volume to various suppliers but executes decisions made by the buyer or the purchasing manager.) Some companies show the suppliers the total volume on the item as well as their share of it. This can serve as an incentive for each of them to do a better job and gain a greater share of the business.

WHY HELP THE SUPPLIER?

Sometimes people question why they should help their suppliers plan their capacity needs. Their attitude is: "It's not our problem." In fact, it is; without providing the supplier with "forecast" information, the supplier will be forced to guess your future needs. The buyer will have to rely on the quality of the supplier's guess rather than on the quality of the planned orders in the MRP II system. In rising markets, customers often find themselves put on allocation when the supplier "guessed wrong."

The issue is not "to forecast or not to forecast," but merely "who forecasts."

Giving better information to suppliers enhances their ability to give customers what they need, when they need it. Many companies using supplier scheduling have found that high-quality information helps mediocre suppliers become good ones, and good ones become excellent. Their philosophy is: "Let's help them to help us. Their problems are our problems."

ENHANCEMENTS TO THE SUPPLIER SCHEDULE— CAPACITY PLANNING

Figure 3.6 is an example of a supplier schedule enhanced by additional capacity planning information. A "capacity unit of measure" has been added to the right of the item description, for use when a customer is buying a variety of different items from a supplier. Let's assume that these items consume varying amounts of capacity in the supplier plant. In this situation, the supplier can't merely add up the requirements in pieces for a given period and get a meaningful measure of workload.

The capacity unit of measure "translates" piece part requirements into capacity requirements meaningful to the supplier. In Figure 3.6, for example, Part #13579 has a capacity value of 2.5. Therefore, the 100-piece requirement in the week of 2/8 translates to a capacity requirement of 250 for that week.

The other items require different amounts of capacity, the units of measure for which are shown. As with the first item, these are used to translate piece part requirements into supplier capacity terms. For example, the 20-piece requirement for Part #24680 in the current week and beyond is shown as 200 units of supplier capacity. Similar calculations are shown for the other items.

The resulting capacity requirements are totaled and shown at the bottom of the schedule. For columns containing more than one week, a weekly average is shown. (This information points out a potential problem with this supplier, which we'll discuss later in this chapter.)

How does the company obtain these capacity units of measure? By asking the suppliers—they know where their bottleneck operations are and the capacity terms in which they think of them. Further, given the necessary education and training about MRP II and supplier scheduling, they will usually be more than glad to provide them. Some typical examples of suppliers' capacity units of measure are shown in Figure 3.7.

Jones Company supplier schedule for: Smith, Inc. – week of 02/01/9X

Supplier #114

Supplier Scheduler: AB
Buyer: CD

Firm zone: first 04 weeks

Material zone: next 06 weeks

------ Requirements ------
FFFFFFFFFFFFFFFFFFFFFFFFFFFFFFFFF MMMMMMMMMMMMMMMMMM

Item #	Descr.	Cap. u/m		Week 2/1 & previous	Week 2/8	Week 2/15	Week 2/22	Week 3/01	Week 3/08	Next 04 Wks	Next 12 Wks
13579	Plate	2.5	Qty:		100				100	100	300
			P.O.#:		B1146						
			Cap.:		250				250	250	750
24680	Panel	10.0	Qty:	20	20	20	20	20	20	80	340
			P.O.#:	B1122	B1146	B1180	B1203				
			Cap.:	200	200	200	200	200	200	800	3400
42457	Tube	1.0	Qty:	300			200			1100	3500
			P.O.#:	B1122			B1203				
			Cap.:	300			200			1100	3500
77543	Frame	3.0	Qty:			40		40			
			P.O.#:			B1180					
			Cap.:			120		120			
Total capacity requirements:				**500**	**450**	**320**	**400**	**320**	**450**	**2150**	**7650**
Weekly average:									**536**	**538**	**638**

Figure 3.6

Some suppliers have required this type of "translation" into their own internal unit of measure for a long time. One example is sheet metal. Often the purchasing company's unit of measure is square feet. The bills of material are expressed that way, and the material is inventoried and issued in square feet. The supplier's unit of measure is pounds, however, and the supplier wants the information in that manner. This translation can be awkward, particularly in Purchasing and Receiving.

The capacity unit of measure on the supplier schedule can ease this situation. In the sheet metal example, the "QTY" row would show square feet (the company's unit of measure) while the "CAP" row would show pounds (the supplier's unit of measure).

This display of information is also useful when future requirements are changing. The furthest right column of requirements, captioned "Next 12 Weeks," shows an average weekly requirement of 638 units per week (in capacity terms). This is substantially in excess of the customer's requirements now and in the recent past. It's also substantially in excess of the supplier's current level of output, which is closely matching current requirements. Again, this type of information should raise a flag to the buyer, the supplier scheduler, and the supplier personnel. Obvious questions arise:

1. Given the supplier's current situation, can he supply the customer at this higher level of volume easily?

2. If not, what changes can be made in his plant to enable him to do so?

3. Is the supplier willing and able to make these changes?

4. If not, can the company "off load" some of this volume to another supplier?

Examples of supplier capacity units of measure		
Commodity	*Supplier Unit of measure*	*Customer Unit of measure*
Castings	Each	Molds, trees
Forgings	Each	Hammer strokes
Sheet metal	Square feet	Pounds
Cartons	Each	Square feet
Liquids	Pints, quarts, gallons	Liters

Figure 3.7

Seeing these kinds of potential problems ahead of time can make a big difference. It can enable buyers and their suppliers to anticipate problems, rather than being forced to routinely react to them. If the buyer can communicate all of the future requirements to the supplier via the supplier schedule, and the supplier can in turn communicate all of the capabilities and limitations of the equipment used to produce the parts, then the two can work out all of the potential problems in the future before they become disasters in the present. In doing so, both sides can look forward to a long, mutually satisfactory relationship.

ENHANCEMENTS TO THE SUPPLIER SCHEDULE—INPUT/OUTPUT CONTROL

In an MRP II system, one of the key elements is feedback, or tracking performance to the plan. For the "inside factory"—one's own shop—the technique used to track capacity performance to plan is called input/output control. Standard hours of actual output for each work center are compared to its planned output. (The same is true for input to each work center, to ensure that the work is actually there.) Deviations of actual performance to plan are calculated. When the cumulative deviation exceeds a predetermined tolerance limit, corrective action is taken.

A similar approach can be taken for the "outside factories," that is, a company's suppliers. Figure 3.8 shows an example of a supplier schedule with supplier performance information added.

The performance information is shown at the bottom, in the area labeled "Performance Summary." In this example, prior weeks' requirements are expressed in capacity terms—"Capacity units required"—as are receipts, so an "apples-to-apples" comparison can be made. The cumulative deviation line represents the running comparison of actual to plan. A cumulative deviation that increases beyond a predetermined tolerance limit would be a signal to the buyer, supplier scheduler, and supplier personnel that corrective action must be taken.

The last line shows deliveries missed by the supplier. This highlights supplier performance to specific deliveries, as opposed to performance to the overall plan.

The information in the two columns farthest to the right on the bottom of Figure 3.8 indicates that the supplier's volume performance year-to-date is very good, over 99 percent. But the supplier has missed 10 percent of the planned deliveries, which would normally be considered unacceptable.

Jones Company supplier for: Smith, Inc. – week of 02/01/9X

Supplier #114

Supplier Scheduler: AB
Buyer: CD

Firm Zone: first 04 weeks   Material Zone: next 06 weeks

------- Requirements -------

*FFFFFFFFFFFFFFFFFFFFFFFFFFFFFF*MMMMMMMMMMMMMMMM

Item #	Descr.	Cap. u/m		Current + Past due	Week 2/8	Week 2/15	Week 2/22	Week 3/01	Week 3/08	Next 04 wks	Next 12 wks
13579	Plate	2.5	Qty:		100				100	100	300
			P.O.#:		B1146						
			Cap.:		250				250	250	750
24680	Panel	10.0	Qty:	20	20	20	20	20	20	80	340
			P.O.#:	B1122	B1146	B1180	B1203				
			Cap.:	200	200	200	200	200	200	800	3400
42457	Tube	1.0	Qty:	300			200			1100	3500
			P.O.#:	B1122			B1203				
			Cap.:	300			200			1100	3500
77543	Frame	3.0	Qty:			40		40			
			P.O.#:			B1180					
			Cap.:			120		120			
	Total capacity requirements:			500	450	320	400	320	450	2150	7650
	Weekly average:								536	538	638

Performance summary-prior wks.	Last wk	2 wks ago	3 wks ago	4 wks ago	4-wk tot.	*Ytd.	Volume	%
Capacity units required:	580	495	520	565	2160		22427	
Received:	582	580	406	573	2141		22298	
Cumulative deviation:	2	87	-27	-19	-19		-129	-0.1
Deliveries missed:	0	0	1	0	1			10.0

Figure 3.8

Consider the similarities between information for the "inside factory" (the company's plant) and the "outside factory" (their suppliers' plants):

1. With MRP II, companies can give their plant realistic, valid, attainable schedules of what to produce and when. They can do the same thing for their suppliers.

2. With MRP II, companies can do capacity requirements planning for their own plants. They can also do it for their suppliers.

3. With MRP II, companies can track input/output performance for their own plants. They can also do it for their suppliers.

4. Since the schedules are valid, the plants operate on the principle of "Silence is approval." This means that, as long as they will hit the schedule, nothing needs to be said. When something goes wrong to cause a missed schedule, then they need to give feedback.

In companies doing supplier scheduling well, routine "follow-ups" of suppliers have become a thing of the past. Suppliers know that the due dates are valid. They know the company needs the items at those times. When suppliers have a problem and can't ship on time, they know it's necessary to notify the company at once. Otherwise, they say nothing and ship on time. Silence is approval.

ELECTRONIC DATA INTERCHANGE (EDI)

Two of the hottest issues in purchasing these days are JIT, which allows the buyer to receive parts more frequently, and EDI, which allows companies to do "electronic purchasing" with their suppliers. There are many forms of JIT scheduling (see Chapter 4) and EDI (see Chapter 9) that can give companies and their suppliers instant access to purchasing and scheduling information through either Kanban signals or computer-to-computer linkages. The key operating words for both JIT and EDI are "well managed"; you can take the fanciest JIT or EDI system in the world and obtain terrible results if the schedules are invalid and people misuse the tools. As with anything else in business, the mere presence of the computer will not guarantee excellent results; computers in manufacturing are only as beneficial as the people who use them and the validity of the schedules they have to work with. If you are going to be successful, you need to recognize that fact and develop valid schedules first, then you can approach JIT and EDI in a sensible and realistic manner.

SUMMARY

- Supplier scheduling allows buyers the time to do what they do best—save the company money.

- Paperwork is significantly reduced when supplier scheduling is used, because the need for most hard copy is eliminated.

- Lower inventories, constant pricing, and shorter lead times are other benefits of supplier scheduling.

- Supplier scheduling, done properly, results in a reduced firm future commitment.

- Suppliers' on-time delivery improves significantly.

- While JIT encourages single-sourcing, the reality of the manufacturing world dictates that some multiple-sourcing must still be done. Multiple-sourcing is possible with supplier scheduling.

- Communication between the buyer's company and the supplier's company improves.

- If anyone asks, "Why help the supplier?" just remind them that the supplier's problems are your problems, too.

- Supplier schedules can be enhanced by showing capacity and input/output data.

- Electronic Data Interchange (EDI), which involves computer-computer linkups for "electronic purchasing," is a useful means—but it is not an end in itself.

Chapter Four

Raising the High Bar

In the last chapter, we described supplier scheduling. Companies that implement supplier scheduling achieve dramatic results, including improved credibility with the plant and suppliers, significant reductions in purchased part inventories, a reduction to 1 or 2 percent of orders past due, the elimination of expediting and hot lists, the communication of planning data to the suppliers, a reduction in supplier lead times, and improved service to the plant. In addition, companies that use supplier scheduling find that their buyers, instead of spending time expediting orders, can focus on carrying out value analysis, negotiations, and problem resolution with suppliers. Finally, supplier scheduling can lead to a highly motivated and professional Purchasing Department that has the time and information needed to do the best possible job.

While all these benefits are certainly desirable, they're not the end of the line; purchasing professionals should continuously challenge their level of performance today and strive for improvements tomorrow. That means asking questions like: Is there something else we can do to further reduce lead times, inventory levels, and lot sizes, while providing better service to the shop? Is there anything else we can do to eliminate more waste—things that add cost but not value to the product?

A number of areas offer excellent potential for achieving such improved performance in purchasing: Kanban, supplier setup reductions, design for manufacturability, transportation issues, supplier assurance of quality, and electronic data interchange. These areas will be discussed below and in the following chapters. First, though, we'll cover the JIT/TQC process, in order to have a basis for understanding how JIT/TQC drives continuous improvement.

The JIT/TQC Process

The philosophy behind JIT/TQC is the elimination of waste, waste being defined as anything that does not add value to the product. In purchasing, there are many kinds of waste that are encountered every day. We receive requisitions from planning, we issue purchase orders and distribute the copies to several departments and the supplier, we write up receiving reports and get invoices for payment. None of those pieces of paper add value to the product, only cost. Purchased material is received by the Receiving Department, where the paperwork is checked and the quantity of the items verified. Those parts are then moved to Quality Control, where the items are checked against the specification. The material is then moved to the warehouse, where it is put away and stored until needed. When needed, the material is then counted out by the warehouse and moved to the manufacturing floor. None of these activities add value to the purchased material, only cost. The ultimate goal of JIT/TQC is the total elimination of waste.

To meet the JIT/TQC goal of zero waste, we must learn how to economically manufacture "one less at a time." (See Figure 4.1.)

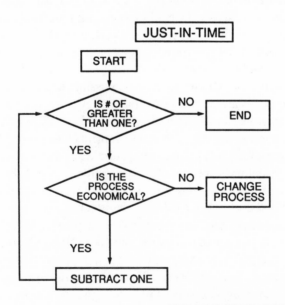

THE JIT JOURNEY BEGINS AND
PROGRESSES ONE LESS AT A TIME

Figure 4.1

Let's say you are manufacturing a product in lots of 500, and that is an economical run quantity for your company. The "one less at a time" flowchart states that if it is an economical lot size, subtract one. Could you manufacture in a lot size of 499? The obvious answer probably is "yes"—it is such a small change. Could you manufacture in lots of 498, 497, 496, etc.? We continue the process down to 350. Is the process economical? It's not as economical and we are having trouble getting the work completed because of the increase in the number of setups. In other words, we have uncovered a constraint.

At this point, the flowchart states to change the process. Try and find out how to lower the setup time. When we get the setup down, we then continue the process. Could we make 349, 348, 347, etc.? We get down to 250 and the material-handling process begins to become a constraint. We have increased the activity in the stockroom and the material handling on the plant floor. These costs are starting to go up, not down. The answer lies in changing the material-handling process. Maybe we can rearrange the equipment on the plant floor in a cellular flow, store the material at point-of-use on the plant floor, and greatly reduce the need for material handling.

When we reduce these costs, we again continue the process. Could we make 249, 248, 247, etc.? We get down to 125. Now we find out that we have four times the number of shop orders, purchase orders, and related transactions. The workload on Receiving, Quality Control, and the stockroom is now much greater. So the next step is to certify the supplier's quality, have the supplier deliver the material directly to point-of-use on the shop floor, have the personnel on the shop floor use a barcode reader to do the receipt, and eliminate the need for those departments. How about 124, 123, 122, etc.? The process continues until we eliminate all waste in the system.

Note that each of these changes are small steps, and the process is under control the whole time. They are not radical changes. Companies that go from lot sizes of 500 to lot sizes of one overnight usually fail. Why? Because they do not have the time or experience to address the problems that come up along the way. They end up adding costs, not minimizing them, and eventually go back to the larger lot sizes. They say, "JIT does not work for us." JIT works, they just used the wrong process to implement it. And it is not just one less in the lot size. It is one less purchase order, one less piece of paper, one less in inventory, one less inspection in Quality Control, one less defective part, one less receipt in Receiving, one less supplier, one less move of material from

Receiving to the warehouse, one less move of material from the warehouse to the plant floor, etc. It is one less until all the waste, the non-value-adding activities, is eliminated.

MRP II and JIT Functions

How do MRP II and JIT function together? MRP II is the planning and control system that plans and controls the manufacturing and purchasing environment. MRP II gives us the valid schedules we use in purchasing. JIT, on the other hand, is the process used to change the manufacturing and purchasing environment, to execute those valid schedules quicker, faster, and better. Its goal is to eliminate any and all waste associated with executing these valid schedules.

Kanban

One of the key techniques in the JIT process is Kanban. Kanban, or demand pull as it is sometimes called, is a material movement and queue control technique that takes on the flavor of scheduling. The power of Kanban lies not in the use of the cards or signal but in the ability of the process to maintain a flow of work and defect-free products.

Figure 4.2 depicts a situation in which there are two operators. These operators could be in your own or your supplier's plant. Each is authorized to hold three jobs at his work center (represented by three squares to the right of the operator). If all the squares are filled with work, the operator is not authorized to manufacture any additional material. The

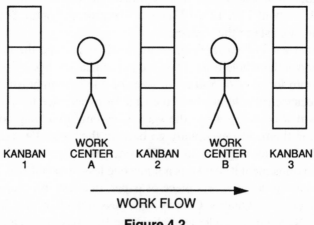

| KANBAN 1 | WORK CENTER A | KANBAN 2 | WORK CENTER B | KANBAN 3 |

WORK FLOW

Figure 4.2

DEMAND-PULL MODEL

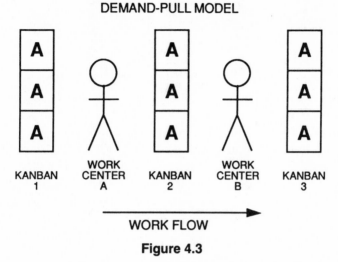

WORK FLOW

Figure 4.3

Kanbans here are represented as squares, but they could also be a card or an inventory location on the manufacturing floor.

The rules of Kanban are fairly simple:

1. Don't start any work without an open Kanban.

2. Work to keep Kanban full.

3. Never pass on a known defect.

Figure 4.3 shows that each Kanban is filled with a job, so the operators are not authorized to do any work.

But at this point, along comes a customer who buys a product from the inventory on the right Kanban. (See Figure 4.4.)

The operator in Work Center B is now authorized to make another item, because there is an open Kanban square to the upper right. The operator pulls some material from the middle Kanban and starts to work on it. This causes the operator in Work Center A to be in demand mode, because there is an open square to the right of Work Center A and that operator begins to work on a purchase part from Kanban 1.

As both of the operators complete their jobs, both Kanban 2 and 3 are full, so the operators are not authorized to start on another job. Note the open square in Kanban 1, which authorizes the supplier to supply more material. This type of Kanban is referred to as "generic Kanban" because only one part was being produced in these work centers.

DEMAND-PULL MODEL

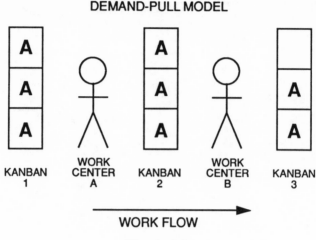

Figure 4.4

Where two or more products are built in the same line, a "brand-name Kanban" is used. (See Figure 4.5.) Note in the example shown in Figure 4.5 that two products are being manufactured, A and B.

When all the Kanbans are full, the operators are in a no-demand mode. Now the customer buys Product B, which authorizes Work Centers A and B to produce another B. When they complete their respective jobs, more purchased parts would be required for B to fill the Kanban square. (See Figure 4.6.)

If the customer had purchased Product A, the open Kanban square would have authorized more purchase parts for A. This works if you have a limited number of part numbers that are used frequently. Where you have a large variety of parts, most companies use a final assembly

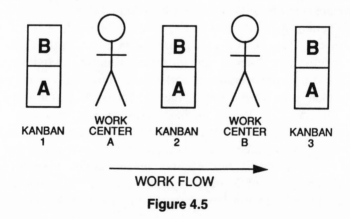

Figure 4.5

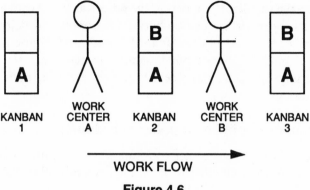

Figure 4.6

schedule that gives the sequence of customer orders to be built. (See Figure 4.7.)

The final assembly schedule details the order in which the product is to be built. In this case, the order is A-A-F-E-A-B-D. The purchase parts would be pulled from A, then F, etc. The open Kanban squares would authorize the supplier to replace the material in that sequence.

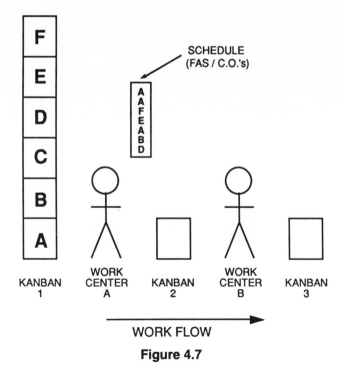

Figure 4.7

Because the supplier is not next to the open Kanban, a card is used to indicate the open square and authorize the supplier to deliver more parts. The card would be similar to that shown in Figure 4.8. This Kanban card indicates which purchase parts to make next, the quantity of the parts to be made, and where to deliver the parts in the buyer's manufacturing facility.

How Kanban Applies to Supplier Scheduling

To see how Kanban fits together with supplier scheduling, consider the following examples. The first is based on a company that makes casement windows for sale in the housing market. The windows are similar to those shown in Figure 4.9 and, for the purpose of this example, are available in twenty different sizes.

The company in this example produces 1,000 windows per week in its manufacturing facility and level loads the factory at 200 windows per day. From MRP, the supplier schedule would show 200 windows per day, and 1,000 windows per week in the future. This authorizes the supplier

SUPPLIER KANBAN CARD	
PART NUMBER	A 173
DESCRIPTION	LEG
SUPPLIER	123
QUANTITY	40
QTY / CONTAINER	20
# CONTAINERS	2
NOTES:	

Figure 4.8

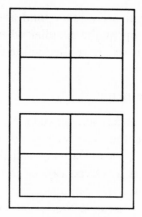

Figure 4.9

to purchase enough glass to deliver 400 pieces per day. (Note: The number is 400, not 200, because there are two pieces of glass per window. The supplier must therefore allocate enough capacity in the shop to cut 400 pieces each day from the master glass sheets.) While the supplier schedule tells the supplier how much product to deliver, the Kanban card specifies the sequence in which product should be made. The result is that purchased parts match the company's assembly operation.

The process works as follows. On Day 1, the manufacturer rough-cuts the wood pieces that will make the window frames from the long master pieces of wood. As the company rough-cuts the lumber, it establishes the sequence by which the windows are to be assembled. This assembly sequence in turn establishes the sequence by which the company needs to receive the glass. As the lumber is cut, a Kanban card is placed in a container to represent the two pieces of glass to be matched with the pieces of wood. At the end of the day, when the supplier delivers glass for the next day's assembly schedule, the supplier's truck would pick up the Kanban cards from the container and return them to the supplier's manufacturing facility.

On Day 2, the window manufacturer routes the desired contour on the window frames, drills all necessary holes for assembly, and cuts out the slot in which the glass will sit. Also on Day 2, the supplier cuts the glass in the same sequence as the Kanban cards and loads the glass into specifically designed carts, also in sequence specified by the cards. At the end of Day 2, the carts are delivered directly to the assembly line at the buyer's shop. On Day 3, the window frames and glass are assembled together in the order both were made, and at the end of the day, the empty carts are picked up by the supplier when he delivers that day's shipment of glass. The carts are refilled in the following days.

Measurable Improvements

Now let's consider the difference between the old way this company purchased its glass and the new way based on Kanban. Before the company used Kanban cards and supplier schedules, it purchased glass in large lot sizes, had it delivered two or three weeks before it was needed in manufacturing, and averaged a month's worth of glass on hand in all twenty sizes. Despite the inventory, the company experienced occasional stock-outs on various sizes. The glass was delivered to Receiving in large heavy wood containers for protection. Receiving would open the containers, count the quantities, and write up a receiving report. The containers would then be carefully moved to Quality Control, where the size and specifications would be checked.

Next, the wood containers would be nailed closed again and moved to the warehouse and placed in special racks in order to minimize breakage during storage. These wooden containers consumed a lot of space in the warehouse, and despite the thickness of the wood, a fair amount of breakage still occurred.

A month later, as the glass was needed, it would be moved in the wood containers and a worker in the factory would stage the glass by size as required for the jobs. Broken pieces would be discarded but would not be charged back to the supplier, because there would be no way to establish who or what caused the breakage. Moreover, breakage was often not recorded correctly, so the inventory was usually incorrect, causing additional unexpected stock-outs.

Since the company implemented a combined approach using JIT and MRP II, in this case using Kanban cards and supplier schedules, it has experienced immediate positive results. Glass is brought in daily and delivered directly to the production line. Any breakage is noted and recorded before the supplier's truck leaves, and the stock is quickly replaced at no charge.

At the end of a day, if the manufacturing facility produces two hundred windows and there is no glass left on the carts, the supplier has to have delivered four hundred sheets that day, and is paid on the basis of the company's schedule completions in the shop. In other words, if the company produced ten windows measuring 60 by 30 inches, the supplier must have supplied the company with twenty pieces of glass measuring 26 by 26 inches.

The benefits to the company, of course, include the elimination of Receiving inspection and Quality Control inspection, as well as the elimination of moves from the warehouse to the factory floor, with their attendant handling of the glass. The breakage in inventory due to poor handling has been dramatically reduced, which in turn has eliminated the need for a month's inventory on hand. And the reduced inventory need has reduced the need for all the warehouse space in which the glass was previously stored. Another benefit is that the buyer has negotiated a price over time for the various glass sizes based on the supplier scheduling requirements.

The supplier, too, has experienced significant benefits, including getting the information they need to plan the glass and cutting capacity well in advance of their needs. In addition, his paperwork is drastically reduced since there are no requisitions, no purchase orders, no receiving reports, and no invoices.

Just-in-Time is defined as the elimination of waste, waste being defined as anything not needed to add value to the product. In this example, the use of Kanban cards and supplier scheduling eliminated a good deal of waste.

Faxban

Where suppliers are more distant, the idea of sending the Kanban card to the supplier may not be practical and may take too long. In this case, buyers have started to fax a copy of the Kanban card to the suppliers. This is usually very easy to set up, and most suppliers already have fax machines.

Here is how it typically works. When an open Kanban occurs for a purchase part, the Kanban card that would authorize the supplier to make more material is sent to Purchasing. Purchasing faxes a copy of the Kanban card to the supplier and then sends the original Kanban card to Receiving. The supplier manufactures the material, attaches the copy of the Kanban card to the material, and ships the material to the buyer's company. Upon receipt of the material, Receiving replaces the copy of the Kanban card with the original and moves the material to the appropriate location on the manufacturing floor. If needed, the copy can be sent to Accounting to authorize payment or it can be discarded if the receipt is entered into the computer.

In some situations, because of the distance the supplier has to ship the material, weekly shipments might be preferable to daily or more frequent shipments. In these cases, the supplier accumulates a week's worth of faxes and then ships all the material at one time. When the entire shipment is received, the original Kanban cards are again attached to the individual lots of material and the material is moved to the appropriate location.

OTHER KANBAN TRIGGERS

In the above example, a card was used to trigger the production of more material, but other items can also serve that purpose. Companies have used golf balls, doughnuts, tubs, and other items instead of a card.

One manufacturer uses a warehouse tub to authorize more work. The tub is made of steel and measures 3 by 4 by 4 feet. The buyer sends one hundred parts to the supplier per tub and the supplier paints the parts, puts them back into the tub, and the buyer's truck picks up the painted parts the next day. The tubs are color coded. The supplier schedule indicates the daily quantity of parts the supplier will be required to paint, while the color-coded tubs indicate the color the parts are to be painted and how many of each color are to be produced.

A furniture manufacturer uses a variety of sizes of wood tops, depending upon the size of the desks to be produced. The standard size is 30 by 60, but tops come in sizes as small as 24 by 48 and as large as 36 by 76. The manufacturer makes up a final assembly schedule similar to the one shown earlier in Figure 4.7. Because of the weight of the wood core used in these tops and the large quantity used each day, the supplier has to make eight deliveries a day to satisfy the buyer's needs for wood core. In this case, the Kanban trigger or authorization for the supplier to produce more parts is a telephone call by the final assembly scheduler at the buyer's plant. Each hour the final assembly scheduler phones the supplier and gives him the sizes and sequence of wood core to be delivered. The supplier then cuts the sizes from master sheets of wood core and sequences per the schedule. Two hours later the wood core is delivered directly to the top line by the supplier. The supplier schedule authorizes the supplier to purchase master 4 by 8, 5 by 10, and 6 by 12 sheets of wood core, and the telephone call (Kanban trigger) tells the supplier how to cut the individual pieces.

Here's another example. A manufacturer of office chairs sells a line of products that offer the customer the option of having the arms powder-coated in any of eight different colors. The manufacturer sets up eight separate racks in the chair assembly area, each one containing different color arms. The arms are put in the racks in pairs of left and right units.

At 7:00 A.M., the supplier goes to the chair assembly area, counts the number of openings in each of the eight racks (an opening represents a left- and right-hand arm pair), and records the number of openings by color. The supplier then walks over to the weld department to a large rack that contains the arm weldments and removes the number of arm pairs needed to fill the powder-coated racks.

Next, the supplier takes the arm assemblies back to the shop and powder-coats the arms that same morning. After the supplier leaves, the weld department counts the number of holes in the arm weldment rack and replaces those with new weldments during the morning. At 1:00 P.M., the supplier delivers the powder-coated arms and puts them in the appropriate colored racks. The powder-coated arm racks are not full at this point because the chair assembly department has been pulling arms during the morning to make chairs.

At this point, the supplier counts the holes in the racks, goes to the arm weldment racks, and pulls arm weldments that are powder-coated during the afternoon and delivered at 7:00 the next morning. The weld department replaces the arm weldments again in the afternoon to ensure that there will be sufficient arms for the supplier at 7:00 the next morning. In this case, the MRP supplier schedule tells the supplier how many arms will be needed daily and authorizes the purchase of the powder in the eight colors; the openings in the racks tell the supplier the actual sequence in which to powder-coat the arms.

When combined with the supplier schedule, a Kanban system in Purchasing can be a powerful tool. Why not pick a supplier who gives you a variety of parts and see if you can develop a Kanban mechanism today? You won't regret it.

TOOLING CHANGEOVERS

One of the biggest obstacles in most companies to reducing order quantities and getting more frequent deliveries is the time it takes to set up the tooling or change over the equipment to run the items. If your supplier takes eight hours to set up the tooling or change over the equipment, he won't get excited about producing and shipping very small shipments to you. Therefore, another area that the buyer will need to work on with the supplier is reducing changeover times.

Let's first discuss how the setup affects the purchase quantity you buy, and then talk about how to reduce setup costs. Most companies use a form of the economic order quantity (EOQ) to calculate the quantity they need to produce. Figure 4.10 graphically demonstrates how the EOQ calculation works.

Line A represents the fact that the total cost of the material increases incrementally as we produce a larger lot size. A costs $.10 each. If you produce one, the total material is $.10; if you produce 10, the total cost of the material is $1.00; if you produce 100, the total cost of the material is $10.00, and so on. Line B represents the cost of the setup allocated to each piece. If the setup costs $100.00 to complete and you produce ten parts, the cost of the setup is $10.00 per piece. If you produce 1,000 parts, the cost of that same setup is $.10 per piece. If you produce a lot of 10 pieces, the material cost is $1.00 (10 pieces times $.10 each), and the setup cost is $10.00 per piece. The sum of A and B is $11.00. At a quantity of 30, that cost drops to $6.33, and by the time you get up to 50, the cost is back up to $7.00 each.

Therefore, the lowest cost to produce that part is with an EOQ of between 30 and 35 units. If you could lower the setup or changeover cost to $50.00, the EOQ would drop to between 20 and 25 units. The lower

ORDERING COSTS

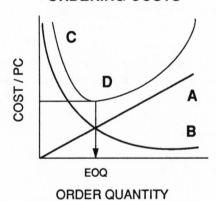

ORDER QUANTITY

Figure 4.10

where A = cost per piece times the quantity
     B = setup cost divided by quantity
     C = sum of lines A and B for each quantity
     D = EOQ quantity or the lowest cost to produce the item

the setup cost, the lower the run quantity. Ideally, if you could lower the setup far enough, the run quantity would become one.

A major lift truck manufacturer in the United States had a series of six hundred-ton presses in the shop. The average setup time per tool was eight hours. The Japanese competition also had the same six hundred-ton presses, but they had reduced the same setup time to three minutes. The U.S. manufacturer had to produce a large quantity to justify the setup, while the Japanese company produced daily what they needed. The U.S. manufacturer was not able to get product out of the machine while the tool setup was in progress, while the Japanese company was able to produce parts all day long. Guess who had the competitive advantage in the marketplace? Purchasing has to work with its suppliers to reduce setup times so the buyer can purchase smaller, more frequent shipments at a lower cost. That means buyers need to understand the techniques to reduce changeover times.

One of the key techniques in setup reduction is to videotape the setup process and put the operators and setup people in a room to watch the video. Have them break the elements of time into types of activity, and then discuss which activities can be eliminated or reduced. One company did this and generated a rather entertaining video. People chuckled as they watched a lift truck remove the tool currently in the press, after which the lift truck operator drove to the tool room to exchange it for the next tool to be used. The next tool was not available (it was in Maintenance), so the lift truck operator drove all the way back to the machine to ask, "Which tool do you want as an alternative?" This took around half an hour. What were all the setup people and the operator doing during this time? Standing around waiting for the new tool.

The solution this company used was to put a two-way radio on the lift truck. Before the lift truck operator left the press area, he called and verified that the tool was available. While he drove to the tool room, the tool was picked and staged so the lift truck operator had no waiting time. The average setup was reduced by twenty minutes as a result.

The second technique is to separate those internal and external activities. An internal activity is one that requires you to lock out the power so you can't produce parts while the activity is occurring. An external activity is one that allows you to produce parts while it is occurring. In the preceding example, the lift truck operator could have gotten the new tool while the last production run was still going and brought it to the machine prior to starting the change. The object is to move as many activities to external status as possible.

One company purchased a few extra beds for the press they owned. This allowed them to completely set up the next tool outside the machine on one of the extra press beds, and then, by means of a series of roller tables, slide the new tool into the press and use a quick connect system for securing the tool. This cut the setup time from eight hours to fifteen seconds! The company only runs the parts it needs each day now, not large lot sizes to justify the eight-hour setup.

Finally, let's take the example of changing the tire on your car to illustrate how you can improve the process. If you have a flat tire, you first need to open the trunk, get out the tire iron, and remove the hubcap. Second, you must get the jack and put it under the car. You then loosen the five lug nuts, jack the car up the rest of the way, and remove the five lug nuts. Next, you take the new tire out of the trunk and reverse the process to put it on. Assuming you can find all the parts of the jack and that the spare tire has air in it, this process will take you around fifteen minutes.

Now consider an Indianapolis 500 race car. Where is the jack? Built into the car. How many lug nuts? One, and it is mistake-proof so you can't put it on wrong or misaligned. Changeover time: fourteen seconds. An Indianapolis 500 car going two hundred miles an hour travels a hundred and five yards in a second. If you delay your tire changeover four seconds, you are a quarter of a mile behind the competition. We need to challenge our suppliers (and teach them when necessary) to lower setup times. If they are successful, then we can get JIT deliveries weekly, daily, hourly, depending upon our needs.

DESIGN FOR QUALITY AND MANUFACTURABILITY

One of the key elements of JIT is in the initial design of a part. A company needs to make sure that the purchase part is designed so that the supplier can provide consistent quality parts at the lowest possible cost. The simpler the design of the part, the fewer parts or processes needed, the better chance the supplier will have of making a quality part every time.

JIT philosophy says, "If you need a part, excel at producing it. Simplify the part and mistake-proof the process so you can't make it wrong." JIT philosophy also says, "If you don't need a part or a process step in making the part, eliminate it. It only adds cost, not value."

When we say mistake-proof (the term used for mistake-proofing is POKE-A-YOKE), think of a jigsaw puzzle; the parts can only go to-

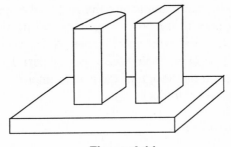

Figure 4.11

gether one way. You can't put it together incorrectly. Figure 4.11 is an example of this.

Two parts need to be inserted into the plate. One is half round with a flat side. It can only be inserted one way, so that perfect alignment is guaranteed. The second part has a different configuration, rectangular, so it can't be inserted in the wrong slot. With these simple design features, we just accomplished two things. First, we have guaranteed the quality of the assembly because you can only assemble the three parts one way—the correct way. Second, you have minimized the operator training necessary because the shape of the parts tells you how to assemble the product.

Another example of mistake proofing is the elimination of fasteners. Fasteners can be time-consuming to insert, can sometimes be difficult to align, and can back out if not tightened correctly. If the part can be designed to snap or slide together, the fasteners can be eliminated.

A manufacturer of copy machines made a major effort to eliminate the use of fasteners. It was able to reduce the number from 130 to one, but couldn't seem to eliminate the last screw. The company offered a $5,000 reward to the first employee who could find a way to eliminate the last fastener. The fastener quickly disappeared from the design. All parts are now modular, and snap together with mistake-proof locking devices that can't come apart under normal usage and can only be put together in the correct sequence. As a result, the manufacturer has reduced the number and cost of the purchase parts, the assembly costs, the indirect costs associated with ordering and inventorying all the fasteners, and at the same time they have significantly reduced the quality problems associated with incorrect assembly of the copy machines.

JIT companies involve the supplier in the design of the parts. How often has a company designed a part only to find it nearly impossible to find a supplier who can manufacture the part to the specification? By involving the supplier in the process, you can assure yourself that the part can be manufactured to the specification. Also, because the supplier is typically an expert in producing that type of a part, he may be able to make suggestions on the design of the part to improve the quality and reduce the cost of your product.

ONGOING IMPROVEMENT

As described earlier, JIT defines "waste" as anything that doesn't add value to the product. Paperwork, poorly designed parts, and setup times do not add value, only cost. Challenge these areas, lower or eliminate the costs associated with them, and you will enjoy solid benefits from putting JIT techniques to work in your company.

As we move to more frequent shipments—weekly, or possibly daily—two areas become critical. The first is transportation costs, which can escalate dramatically with smaller, more frequent shipment, if not managed correctly. The second is quality—essential in a JIT environment because there will be little or no inventory to serve as a hedge against rejections. Both issues will be covered in the next two chapters.

SUMMARY

- Purchasing professionals should continuously challenge their level of performance and strive for improvements.
- MRP II plans and controls the purchasing environment while JIT is the process to change that environment.
- The philosophy behind JIT/TQC is the elimination of waste.
- The "one less at a time" process is a key technique in the elimination of waste.
- Another key technique in the JIT process is Kanban, a material-movement and queue-control method.
- When companies have a combined approach using both supplier schedules and Kanban cards, they experience very positive results.
- One of the biggest obstacles to reducing order quantities and getting more frequent deliveries is the time it takes suppliers to set up their tooling and equipment.
- JIT companies involve their suppliers in the design of the items they buy in order to improve the quality and reduce the cost of their products.

Chapter Five

Organizing for Supplier Scheduling

The goal of supplier scheduling is to get the buyer out of the traditional mode of using paperwork and expediting into one in which he or she has the time to spend money well. The buyer's full-time job becomes one of value analysis, negotiations, supplier selection, alternate sourcing, quality negotiation, lead-time negotiation, integrating suppliers into the business, and developing long-term partnerships with the suppliers. In terms of the day-to-day buying, the buyer should be involved in the exceptions—those things that do not go according to plan. To do this, changes in the purchasing organization are necessary. Figure 5.1 shows the evolution that occurs in most companies as they evolve from a traditional organization of buying to an MRP II and then JIT organization.

Organization

Traditional	Planner writes requisitions	Buyer places orders
Combined	Buyer/Planner plans and also orders	
Supplier Scheduler	Supplier scheduler plans and orders	Buyer negotiates contracts, solves major problems, etc.

Figure 5.1

THE TRADITIONAL METHOD REVISITED

As we saw earlier, most companies have a planner who determines the needs for purchased items. This person enters the quantity and due date on a requisition and forwards it to Purchasing. Upon receipt of the requisition, barring any difficulties, the buyer creates the purchase order.

Unfortunately, in many companies, regardless of size, the planner and buyer do not work in concert. The reason is simple. The planners are normally held accountable for the dollars in inventory. Therefore, they tend to requisition smaller quantities to keep the inventory low. In contrast, the buyers are accountable for price, so they tend to work at getting the largest quantity price break possible. If they're unable to get the maximum price break, they may go back to the planners and ask them to increase the order quantity. At this point, the company has unwittingly set conflicting goals for both people.

Another common problem with the traditional approach is the lack of a valid planning and scheduling system. Without valid schedules, the initial due dates on requisitions may be incorrect because they may not reflect when the material is *really* needed in the plant. Also, because of constantly changing priorities in the factory, it can be very difficult for the due dates to be kept valid during the life of an order.

In such an environment, the buyer doesn't really know when the orders are needed. And if the company runs out of a material before the next shipment from the supplier arrives, destructive finger-pointing begins. The planner blames the buyer for not respecting the dates on his requisitions, and the buyer fires off a return volley by accusing the planner of giving incorrect need dates.

Yet another problem with the traditional method is that in most companies, planners are not allowed to talk to the suppliers. The reasons for such policies are several-fold. First, purchasing people feel they don't get valid dates on the requisitions from the planners, so they don't want the planners passing the unrealistic dates on to the suppliers.

Many purchasing people also fear that planners may treat the suppliers unfairly, violating lead times or perhaps minimum order quantities. In other words, they're concerned that planners may break down the rapport the buyers have worked so hard to establish with their suppliers. Finally, since the planners in most companies are organized by product lines, not commodities, two or more planners could end up contacting a supplier at the same time and giving the supplier conflicting priorities.

THE COMBINED METHOD (MRP)

As industry made the transition from manual to computerized inventory record-keeping, a number of opportunities opened up for purchasing professionals. As the record-keeping became less time-consuming, it became practical to allow one person, typically a buyer, to do both the buying and the planning. That person is often called a buyer/planner.

With this combined form of organization, buyer/planners receive computer-generated information on each of the purchased items. They review this output, reacting to the action messages on each item as needed. Once the planning is complete, they place the purchase orders with the supplier and perform all the normal purchasing functions.

This approach gives buyer/planners much better information than the traditional method. They can see the need dates on every item and can easily increase order quantities or combine items to get the desired price breaks. They're also freed from requisition paperwork, so their overall paperwork burden is reduced. Further, with an effective MRP II system in place, the dates can be kept valid, thereby sharply reducing the buyer's expediting workload.

Despite the benefits, though, the combined approach does pose some drawbacks. The buyers now have to do the planning in addition to their regular buying responsibilities. Some, possibly much, of the time gained by doing less paperwork and having better dates now must be devoted to the planning function. Once again, buyers may be confronted with the classic problem of having insufficient time to do the really important parts of their job: sourcing, negotiation, value analysis, etc.

THE SUPPLIER SCHEDULER METHOD (MRP II)

In today's manufacturing environment, supported by MRP II, there is a better purchasing approach. Since MRP II can generate and maintain valid need dates, much of the traditional conflict between the buyer and planner can be eliminated. Planners can be organized by commodity and can work directly with the supplier on details of the schedule. Often, they're given a new title: supplier scheduler. The buyer's job now changes to what it should have been all along: spending money well.

Note that it's not necessary to go through the combined method to get to supplier scheduling. A company should go right to supplier scheduling from whatever its current arrangement may be.

The Supplier Scheduler's Job

The supplier schedulers must carry out a number of important activities. First, they analyze the Material Requirements Planning reports and react to the messages out of the system. The types of messages they would review include:

- Release an order.

- Reschedule an order in.

- Reschedule an order out.

- Cancel an order.

- Resolve data problems.

Material Requirements Planning looks out into the future and calculates when the company will need to receive more material from its suppliers on each purchased item. It takes the date the company will need more material, backs up by the supplier's lead time, and then recommends when to place an order. This is called a "planned order" because it is only a recommendation from the system. For example, as time passes, a planned order release moves from Week 4 to Week 3, from Week 3 to Week 2, and from Week 2 to Week 1. When it gets into Week 1, the Material Requirements Planning system gives the supplier scheduler a message to firm up the order with the supplier. The supplier scheduler then converts that planned order to a "scheduled receipt," which authorizes the selected supplier to produce the item. If there is a problem from the supplier's viewpoint, the buyer may need to get involved and resolve the difficulty prior to the placement of the order.

Material Requirements Planning is also constantly reviewing the need date (when the first piece is needed) against the due date (when the supplier has been asked to deliver the item). If the need date changes, either in or out, the system gives the supplier scheduler a message to move the due date accordingly (reschedule in or reschedule out).

In either case, the supplier scheduler contacts the supplier and requests the change in due date. If the supplier agrees to the change, there's no problem. The scheduled receipt's due date on the computer is simply changed to reflect the new date. If the supplier cannot or will not change the date, the supplier scheduler may contact the buyer, who tries to resolve the differences.

On occasion, the system will also recommend canceling an order with a supplier. This can happen when a customer has canceled an order and the manufacturer no longer needs a scheduled purchased item. In such cases, the supplier scheduler contacts the supplier to try to stop the order. If there are problems, and perhaps cost implications, the buyer would need to get involved and negotiate the necessary settlement.

MRP II will also generate messages to indicate data problems. A good example of this would be the supplier's weekly capacity. Consider a supplier who works five days a week, eight hours a day, on one purchased item. In a typical week, the supplier produces ten thousand parts. This maximum of ten thousand parts would therefore be input into the system.

Now suppose business picks up and the manufacturer needs eleven thousand of that item in a particular week. The system would give the planner a message that says, in effect, "This exceeds the supplier's stated capacity." The planner would contact the supplier and either get the supplier to produce the larger quantity or perhaps pull part of the eleven thousand quantity into the previous week. If this is a problem for the supplier or a cost is involved, the supplier scheduler would once again involve the buyer to resolve the situation.

In short, the supplier scheduler does all the planning via MRP II, handling the 90 to 95 percent of the items that are routine. The few exceptions are handled by the buyer. Freed of all the traditional paperwork and routine replenishment planning, buyers now have the time and the opportunity to do their job of purchasing better.

The second area of responsibility of the supplier scheduler is to maintain the information in the system. If a price changes or the buyer negotiates a new lead time or selects a new supplier, the buyer passes the information on to the supplier scheduler, who inputs it into the system. After verifying that the master data are correct, the supplier scheduler lets the buyer know that the change has been completed. This allows buyers to exercise management control over the Material Requirements Planning system, but frees them of actually having to do the time-consuming transactions themselves.

The third area of responsibility of the supplier scheduler is to respond to changes to the schedule, to engineering changes, and to supplier delivery problems. "Can the supplier move the shipment of parts from Thursday to Monday?" "Can they give us the red ones before the blue ones?" "We just broke the tool. Tell the supplier we won't need the half-inch size until next week. What size can they move up sooner?" The supplier scheduler thus becomes the eyes and ears of both the supplier and the inside shop.

The last area of responsibility of the supplier scheduler concerns the actual generation of the supplier schedules. After all the necessary planned orders have been converted into scheduled receipts in the Material Requirements Planning system, all the reschedule messages completed, and all the changes updated, the system can then be authorized to generate the supplier schedule. The supplier scheduler is responsible for transmitting the weekly schedule to the supplier. Other responsibilities may include checking with the supplier to ensure receipt of the schedule and to verify that the schedule is practical and attainable.

THE BUYER'S JOB

The supplier scheduler has freed the buyer of the paperwork associated with requisitions and purchase orders and taken on the bulk of the day-to-day contact with the supplier. As a result, buyers can focus on spending money well, which is what they were hired to do in the first place. The buyer's full-time job becomes value analysis, negotiation, supplier selection, alternate-sourcing, quality negotiation, lead-time negotiation, and developing long-term supplier partnerships. Buyers now have time to sit down with members of the engineering team to understand the function of an item and its critical elements (tolerances, specifications, etc.). They also have new opportunities to work with suppliers to find ways to satisfy the function of the item at a lower cost. Finally, there's time to do effective negotiation on all the items, not just the high-ticket ones.

Appendix B details a survey of successful users of Manufacturing Resource Planning, focusing on their results in purchasing. In this survey, the companies using supplier schedulers reported an average 13 percent annual purchase cost reduction. We believe this is primarily because the buyers had the time to do their jobs (in addition, of course, to furnishing the suppliers with valid schedules generated by MRP). Those companies that have buyer/planners reported an average 7 percent annual cost reduction. Why the difference? Under the buyer/planner concept, the buyer has to spend time doing the planning and follow-up function. That takes time away from value analysis, negotiation, etc. With supplier scheduling, the buyer's full-time job is getting the best return on the money he has to spend.

Buyers also have the time now for supplier selection. As we'll discuss in Chapter 12, they can use supplier measurement reports to evaluate how well the supplier is performing. There's also time to search out alternative sources, either to assure supply or to replace a supplier who isn't performing up to expectations. In addition, buyers have the opportunity to negotiate quality and resolve differences between the supplier's specifications and their own. Quality measurement reports identify the parts below quality standards, and buyers have time to then work with the suppliers to install statistical process quality control. Finally, buyers have time to work with Engineering on tolerances, and to help Engineering with standardization programs so as to improve incoming quality.

When supplier scheduling is in place, buyers have the opportunity to negotiate lead times more easily. By visiting the supplier's plants, buyers can determine the real elements of lead times from the supplier's perspective. They can get answers to such questions as: When does the supplier need to order raw materials? When does the supplier set aside capacity to make the parts? How long does it actually take to produce the item? By using the supplier schedule to forecast raw material and capacity needs, the buyer can negotiate a shorter lead time based on the supplier's actual manufacturing time.

In short, buyers are given the time to do all the functions of their job well. Since they're now an expert at buying, rather than at doing paperwork and carrying out expediting, they can become highly motivated professionals in the Purchasing Department.

A POTENTIAL DILEMMA

To whom should the supplier schedulers report? Purchasing managers may feel that the supplier schedulers should report to them, since they're in direct contact with the suppliers. Managers of production and inventory control might want the supplier schedulers reporting to them, since they're operating MRP. What's the solution to this potential dilemma?

The answer lies in the basic issue of accountability. We believe that the supplier scheduler should report to the manager who has accountability for the performance of the purchased material inventory. Inventory performance in this context has two aspects: service and turnover.

Service refers to manufacturing and the customers. This means keeping Manufacturing supplied with purchased items so it can operate efficiently and without interruption. It also means providing good customer service on purchased spares and finished goods. Turnover simply means keeping the inventory levels low.

The manager who can best be held accountable for purchased inventory performance is the logical choice to manage the supplier scheduling operation. In some companies, this might be the purchasing manager. In others the production control manager, or perhaps the materials manager, might be the best choice.

All things being equal, it is important that the supplier scheduling group be a part of the Purchasing Department. In this arrangement, there's only one person to consult if there is a problem with a purchased item—the purchasing manager. It makes no difference if the item was ordered too late, if the supplier didn't ship on time, if the quantity is wrong, or if the inventory level is incorrect on a purchased item; the purchasing manager is ultimately responsible for all purchased item problems. If the supplier scheduler reports to Production and Inventory Control, there may be confusion on whom to call, the purchasing manager or the production and inventory control manager.

Also, if the supplier scheduler reports to Production and Inventory Control, there may remain some conflict between order quantity and price breaks. If the supplier scheduler reports to Purchasing, the purchasing manager is held accountable to manage pricing while staying below the total inventory dollar limit.

That's the theory. In actual practice, though, there are no absolutes regarding the issue of reporting. In the survey mentioned earlier (see Appendix B), about half of the companies had the supplier schedulers as part of the Purchasing Department, and roughly half had them in Production and Inventory Control.

In most companies, regardless of their reporting structure, the supplier schedulers sit in close proximity to the Purchasing Department. This makes it easy for supplier schedulers to update buyers on potential problems as they occur. In return, the buyers keep the supplier schedulers informed on key issues.

Quality problems can be handled jointly by the buyers and supplier schedulers. The supplier schedulers need to get a replacement shipment, while the buyers need to resolve the underlying reason for the quality problem. The buyers also need to keep the supplier schedulers up to date on pricing and lot size changes so they can input the data and replan as necessary. The supplier schedulers keep the buyers updated on engineering changes and new product timing so they can select suppliers and negotiate pricing prior to the effectivity date.

The buyers inform the supplier schedulers about pending changes in suppliers, new suppliers, etc. This can enable supplier schedulers to help new suppliers with material and capacity planning information, and to begin building the core relationship.

In short, buyers and supplier schedulers work together to resolve problems and handle changes before they can cause a production shortage. When suppliers call to discuss quality or pricing, they see the buyer. When they call to discuss delivery, they see the supplier scheduler. But because buyers and supplier schedulers work as a team, they often both meet with suppliers to be sure all bases are covered. Anything short of a team effort may cause a return to the old "your supplier, your schedule" finger-pointing routine, hot lists, and low performance.

How many supplier schedulers will a company need? Our survey showed an average of one supplier scheduler supporting two buyers. Since the buyers and supplier schedulers are set up on a commodity basis, it is fairly easy to arrange the commodities so this 2:1 ratio can work smoothly.

A logical question at this point is "Will we need to hire more people in order to staff the supplier scheduling function?" The answer is almost always "no." The supplier schedulers can often be drawn from the existing planner group as a result of reassigning planning responsibilities from a product orientation to a commodity orientation on the purchased items.

MATERIALS MANAGEMENT

In many companies today, Purchasing, Production, Inventory Control, and other departments report to a person called the materials manager. This person's job is to coordinate and resolve the problems that can occur between the various materials departments. Purchasing wants larger lot sizes. Production and Inventory Control wants smaller lot sizes. Traffic wants truckload shipments. The warehouse wants even flow of material, etc.

With MRP II, the company now has valid priorities, and many of the problems encountered with the informal system vanish. Hot lists disappear, expediting is dramatically reduced, inventory is lowered, weekly or smaller shipments of purchased items even the flow of material.

Traffic is provided with advanced schedules of inbound freight with which to negotiate freight rates.

Consequently, much of the need for a "referee" is removed. The supplier schedulers are given the responsibility of scheduling purchased items and staying within their inventory target, and the production schedulers are given the responsibility of scheduling work-in-process and staying within their inventory target. Therefore, in many MRP II companies, the need for the materials management form of organization is lessened.

MULTI-PLANT OPERATIONS

Many companies today have multiple-plant operations. They often have a Central Purchasing Department, but their manufacturing operations are decentralized. Typically, each operation does its own production scheduling and generates its own material requirements, then forwards them to a central buying group to be purchased. Normally, they would come to Purchasing on a requisition of some type, rather than as a Material Requirements Planning output, which makes it impossible for Purchasing to do supplier scheduling.

There are two approaches to this problem. If each plant has its own computer and its own MRP II software, the supplier scheduling function can be located at the plant site. Central Purchasing can be supplied with the annual quantities of all the parts used at all the facilities and would do annual contracting, based upon the total combined requirements for each item.

Once the supplier is selected, and the terms or conditions negotiated, the supplier scheduling function at the plant site would do all of the supplier scheduling. Though they would be part of the plant staff, they might have a "dotted-line" responsibility to Central Purchasing in regard to supplier selection, order quantities, lead times, etc.

In cases where the company has a central computer and only one MRP II software package, and the central data processing group does all the Material Requirements Planning processing, Purchasing would have the option to do supplier scheduling at the central location or the plant site. Since all plant requirements are run on the same software, it would be very easy to get the combined requirements for every item for every location on one Material Requirements Planning report.

This would let buyers give each supplier a supplier schedule that showed the total requirements for each item, just as if there were only a single location. This would also be far simpler for the supplier than receiving five different plans from five different plants. At the same time, the supplier schedule would break down the total weekly quantity by plant location and tell the supplier how much to ship to each plant on a weekly basis.

Regardless of which approach is employed, it may be advantageous to generate one corporate supplier schedule in situations where there are common items used by several plants. This schedule would display the total requirements for that item, along with shipping and delivery information.

Some companies use a combination of approaches. Large-volume multiple-plant items are handled by a central scheduling group. A plant scheduling group handles items that are unique to a plant and purchased at a nearby supplier. In cases where some items are purchased overseas by someone located in an overseas buying office, those items can be scheduled from that location. MRP II can provide a great deal of flexibility in establishing "who does what and where." Chapter 11, which covers special situations, presents more examples of organizational flexibility.

SAMPLE JOB DESCRIPTIONS

Below you will find sample job descriptions for a supplier scheduler and a supervisor of supplier scheduling. Obviously, the descriptions would have to be tailored to meet the requirements of each company, but they do give an overview of the general functions related to each job.

JOB DESCRIPTION
JOB TITLE: SUPPLIER SCHEDULER

Job Function

Responsible for the planning and maintenance of the Material Requirements Planning system as it affects purchase parts. Monitors changes in the Requirements Planning system and notifies buyers of changed requirements. Works with suppliers on irregular orders to check on the status of current or future shipments of purchase parts as required.

Summary of Duties and Responsibilities

1. Records additions, deletions, and changes to the item master file and the material requirements file for purchase items.

2. Analyzes the Material Requirements Planning system and advises both the buyer and the supplier of irregularities or unusual demand.

3. Releases new orders to suppliers confirming delivery dates on the supplier schedule. This is accomplished by converting planned orders to scheduled receipts (or firm planned orders if quantities or dates are modified) in the Material Requirements Planning system.

4. Reschedules delivery dates as required by the Material Requirements Planning system and, if the change is inside the supplier's lead time, contacts the supplier before making the change to assure it is practical and possible.

5. Contacts suppliers on matters of current and past due shipments. Works closely with suppliers and production schedulers to assure that the master schedule is workable and valid.

6. Responds to routine questions from suppliers and inside employees via telephone.

7. Other duties as required by supervisor.

JOB DESCRIPTION
JOB TITLE: SUPERVISOR, SUPPLIER SCHEDULING

Job Function

To supervise the supplier scheduling function.

Summary of Duties and Responsibilities

1. Responsible for the purchase part inventory level.

2. Responsible for working with data processing to ensure the Purchasing Department of timely delivery of necessary computer runs, and for requesting and securing all "on request" management information reports, such as commodity buying reports, dollars purchased by supplier report, etc.

3. Responsible for ensuring that supplier schedules are provided to the suppliers on a timely basis (via mail, electronic data interchange, fax, etc.).

4. Supports the supplier schedulers in their contacts with suppliers on matters of current and past due shipments. Contacts with suppliers require tact and discretion, as the company image is at stake.

5. Responsible for providing long-range planning reports to suppliers to do capacity and raw material planning and to buyers to do negotiating and contracting.

6. Monitors flow of work through contacts with other departments, engineering change notices, etc.

7. Coordinates new product effort, as that affects new purchase parts entering system.

8. Works on special projects as directed by purchasing manager.

One caution in regards to supplier schedulers should be kept in mind: The job is critical in terms of giving the suppliers a valid schedule. If the job is not done properly, the buyer will be forced to do a lot of expediting and get involved in problems that will pull him or her away from the primary task of spending the company's money well. Therefore, it takes a skilled individual to do the job correctly. If the job is rated as a clerical job and treated as such, the results will probably not be satisfactory. Supplier scheduling is a very important job and must be treated as such.

THE JIT METHOD

In Chapter 4, the Kanban method of scheduling suppliers was covered. To recap, the supplier schedule authorizes the purchase of raw materials and gives the supplier the overall rate at which the items are needed and the allocation of capacity at the supplier's facility. The Kanban mechanism, be it a card, a container, or a telephone call, tells the supplier the actual sequence in which to run the items. In companies with a level, steady, and consistent daily production rate, the supplier schedule is generated infrequently and is used for future capacity planning only. In those cases, the Kanban mechanism is the only trigger to authorize production at the supplier's facility.

Typically in companies using JIT extensively in manufacturing, both the generation of the Kanban card and the sequence in which items will be made are determined by Manufacturing. In such companies it is possible to have the production planner, final assembly scheduler, production supervisor, or the direct laborer schedule the supplier.

In some JIT companies that flow the product in a cellular manufacturing environment, the direct laborers in that cell have been given the responsibility to establish the daily manufacturing schedules. The direct laborers would look at the master schedule to determine what has to be produced, they would then determine the best sequence in which to produce it, and they would also be accountable for requisitioning the material from the warehouse as needed. They would also be responsible for scheduling the supplier, via the Kanban mechanism, for the material needed to complete that daily schedule. In the Kanban examples given in Chapter 4, the direct laborer on the saw at the window company schedules the supplier; in the company that uses color-coded tubs to determine the paint color, the direct labor operator sends the empty tubs to the supplier; and at the chair manufacturer the direct laborer, in effect, by pulling the powder-coated arm out of the rack, schedules the supplier.

The idea of having Manufacturing, and in many companies the direct laborer, schedule the supplier via a Kanban mechanism is a relatively new concept. But we expect it to be the norm for JIT companies in the future.

SUMMARY

- The main problem of the traditional approach to purchasing is that it does not take into account the time when materials are actually needed.

- The "combined approach" puts buying and planning together, which solves the timing problem. But the new planning burden on buyers often prevents them from doing their jobs as effectively as possible.

- With supplier scheduling in place, the planner works directly with suppliers on details of the delivery date. As a result, buyers can focus on what really saves the company money: sourcing, negotiation, value analysis, and related actions.

- With MRP II in place, 90 to 95 percent of the items handled by the supplier scheduler are routine. The buyer handles the rest.

- In addition to generating the actual schedules, the supplier scheduler is responsible for maintaining information in the system, as well as responding to supplier delivery problems, engineering changes, and other situations that might alter schedule dates.

- The person to whom supplier schedulers report will vary by company. The only hard-and-fast rule is that supplier schedulers should not be placed in the position where they are given conflicting directives.

- In JIT companies, the trend today is to let someone in Manufacturing, the production planner, final assembly scheduler, production supervisor, or the direct laborer schedule the supplier via a Kanban mechanism.

Chapter Six

Supplier Quality Assurance

SUPPLIER CERTIFICATION

In a typical company today, purchase orders are placed with suppliers for a certain amount of purchase material. The supplier ships the material to the buyer's company, where it is received at the Receiving Department. Receiving verifies the quantity and item number and writes up a receiving report, which is sent to several different departments. The material is then moved to the Inspection Department, where it is checked against the specifications and is either accepted or rejected. If accepted, it is then moved to the warehouse, stored, and at a later date moved to the Production Department.

If the material is rejected, it is moved to a hold rack in Receiving until the disposition of the material is determined. Each week, a Material Review Board meets to determine what to do with the rejected material. If it is desperately needed in Production and the bad material can be sorted out or reworked, the rejection is waived and the indirect operation of sorting or reworking is completed. The material is then moved to the warehouse and issued to Production as needed. If the rejected material is determined to be unusable, it is returned to the supplier and Purchasing then has to expedite in a replacement shipment of the material. What did all of these indirect activities add to the value of your product? Nothing, only cost!

In the case where the specifications on the shipment were waived and the rejected material was used, a dangerous message was telegraphed to the suppliers: It is okay to ship defective material in the future also. Let's say the specifications are stated as a range: We want a material that is .55 inches thick, but we can accept and use anything that falls between .50 and .60. The material that is waived was .70. We have therefore told the supplier that the true range we are willing to accept is now .50 to .70. The next time the supplier makes a batch at .70, guess what will happen? The supplier will ship it to the company knowing it is out of specification—but also knowing that if the company waived the specifications last time, it will probably waive them again.

To try and prevent this from happening again, the company sorts out the bad material from the shipment and deducts 15 percent of the invoice total for sorting. Many people assume that this sorting charge will deter the supplier in the future. In fact, all this does is tell the supplier that the company will be glad to pay 85 percent of the value for bad material. Now, let's say that the supplier has two buyers for a certain item. Buyer A will always return the items if they are bad, and Buyer B will sort the bad items and give the supplier 85 percent of the value for the items. What is the supplier's disincentive not to ship the bad items to Buyer B? There isn't any.

What is needed is a move toward certification of both the supplier and the material. If that can be achieved, it is possible to expect defect-free quality, frequent and synchronized deliveries to the point-of-use, and the elimination of waste in the Inspection and Receiving departments.

Supplier certification entails three steps. First, suppliers must be qualified. Second, suppliers must be educated in MRP II and JIT/TQC so they understand your expectations and how they can be successful. And third, suppliers must go through the certification process. Let's take a closer look at each step. (Education will be dealt with in depth in Chapter 13.)

QUALIFYING SOURCES

As we stress throughout this book, the recent trend is toward fewer suppliers and single-sourcing of components. This, in turn, makes it even more important to qualify sources. The following seven-step program can help you do that in an organized and professional manner.

Step #1—Get Management Support

Without management support, no qualification program will be successful. To gain that support, we recommend the following strategy. First, state the program's goals and timetables. The key objective is to find and develop single sources of supply capable of delivering defect-free product, on time, to the point-of-use, in frequent shipments—that is cost-effective.

Next, estimate the total cost of poor performance in terms of delivery, quality, etc., and show the effect on the bottom line in black and white. What is the true cost of a line shutdown caused by a material shortage? What is the cost of inspecting, testing, and rejecting poor-quality material? Develop a solid cost/benefit number. Then excite management about the prospect of becoming partners with your suppliers. Provide case studies that demonstrate the results other companies have achieved through similar programs and share those with management. Finally, become a project champion—get the whole company behind the plan by showing them the benefits. Be prepared to sell the concept and to do it enthusiastically.

Step #2—Decide Which Suppliers to Qualify First

If you try to qualify all your suppliers at once, the program will probably fail. A safer approach is to pick the items provided by the worst suppliers, those suppliers who seem least able to hit the specifications, because you'll be able to realize your greatest payback and savings. By either improving the performance of the worst suppliers or replacing them with excellent suppliers, you will quickly validate the program, achieve some quick financial paybacks, and reinforce your efforts throughout the company.

Step #3—Assemble a Cross-Functional Qualification Team

To be effective, qualification must draw on a number of areas of knowledge. No one person can possibly be an expert in all of the categories that will be used to evaluate the suppliers, so your team should be multidisciplinary and include key players from Purchasing, Quality Control, Receiving, Engineering, Finance, Manufacturing, Sales, and Marketing. The purchasing manager should head up the team to give it direction and ensure that the objectives are on target.

Step #4—Agree on Criteria for Evaluating Suppliers

Chapter 2 described nine criteria for a good supplier. Your qualification team can use them as a starting point, then add your own to tailor to your unique situation. For example, you might need FDA approval or have to meet minority hiring requirements.

Next, the team must decide what performance scores your company needs from its suppliers in order to not only stay competitive, but be the leader in your industry. This is a better approach than determining desired performance with respect to how the suppliers are performing today. In other words, determine the absolute performance that suppliers must demonstrate in order to participate in a partnership relationship. This step is sometimes called "benchmarking."

Suppose one of your suppliers was performing only at 70 percent on-time. Don't approach the situation by saying, "We would like to see eighty-five percent by next June." Rather, start with your needs: "We want ninety-nine percent on-time delivery." Period. Otherwise, you will wind up settling for something less than what you need to achieve your position in the marketplace.

The process of benchmarking entails five phases that most companies use to determine the absolute performance levels required.

1. *Planning Phase.* The team decides which items will be benchmarked and who is the industry leader in respect to those items.

2. *Analysis Phase.* The team studies the competitors' strengths and weaknesses with regards to the items to be benchmarked.

3. *Integration Phase.* The team develops and communicates a strategic plan to capitalize on the company's strengths and minimize the weaknesses.

4. *Action Phase.* The team works with suppliers to implement the strategic plan.

5. *Maturity Phase.* Competitive benchmarking becomes a way of life and the company is able to maintain leadership in its industry.

Step #5—Gather and Analyze Your Data

Qualifying suppliers is not just a subjective process—it's very much based on hard numbers. To obtain the necessary numbers, develop a checklist that enables you to compile standardized information about each supplier. Of course, in some cases you might have to ask some special questions of a particular supplier. Even so, the bulk of the data should be as consistent as possible. Use the list of questions in Appendix A.

Step #6—Make an Initial Qualification

Once you run your existing suppliers through your qualification criteria, some will not be interested in being evaluated so rigorously. Some will say they don't want to share or give up anything, that they are the industry leaders (or believe they are). And some will just be beyond education or hope. The few good remaining suppliers will pass the muster in terms of (1) service, delivery, and quality, (2) financial stability, and (3) technical capabilities.

Step #7—Final Qualification

As in most things in life, common sense prevails in the area of qualification. If the best supplier is based in Europe and you are located in Florida, that supplier might not be ideal for frequent deliveries. If your company rates ethics as very important but your supplier has had some ethical problems in the past, you have a bad fit. If two of the suppliers you evaluated for a particular part both rate the same, you may choose to take both of them into the certification phase to determine which performs better.

Once you make your final cuts, you should give feedback to the winners and losers. You will do the losers a great service if you explain why they were not chosen. With this sort of information, they may come around and become the best suppliers during another round of qualifications. The same applies if you have to pull the business from a supplier who initially met your qualification criteria but later could not keep up with your standards. Say why you found the supplier to be deficient—that may help him to tighten his operation and reach the top tier in the future.

SUPPLIER QUALITY EDUCATION

Once the qualification phase is complete, it is necessary to educate the chosen suppliers and provide a vision of what business will be like in an MRP II and JIT/TQC environment. In these sessions, the company should lay out its new expectations for the future regarding quality, delivery, costs, certification, price reductions, teamwork, etc. (See Chapter 13 for a more detailed explanation of the education process.)

CERTIFICATION PROCESS

After the suppliers have been through the education program, certification can begin. To be able to deliver defect-free quality parts, suppliers will have to begin using quality techniques like statistical process control (SPC). Excellent TQC companies train their suppliers in SPC through in-house demonstrations. It is much more effective to show them how SPC works than to tell them about it. Moreover, the supplier realizes how serious you are about TQC because you are practicing it yourself.

First, visit the supplier and verify that the manufacturing processes and tooling that he uses can make a quality item to your specifications. Your quality and engineering people should accompany you and assist you in evaluating the supplier's capabilities. If you determine that the supplier's process is not capable of producing a quality item to your specifications, work with the supplier to change the process, the tooling, or, potentially, your specifications. If you don't visit the supplier's plant, the first time you will find out you have a quality problem is when you reject a shipment.

Next, your company and the supplier need to inspect the items for defects. The goal is to eventually eliminate the need for inspection, but practically speaking, that won't happen for a period of time. In the meantime, both parties need to identify and eliminate the cause of any defects.

The buyer's company should develop an inspection measurement system that identifies defects. Such a measurement system is described in Chapter 12. Likewise, the supplier should set up an inspection process during production. If the supplier runs a lot of a thousand pieces and then inspects them, potentially all thousand could be defective. But if you insist that the supplier inspects the first, twenty-fifth, and fiftieth piece, etc., and that he shuts off the machine if a poor-quality item is produced, you greatly minimize the chances that you will get a defective item. There is no greater waste than taking good material and good labor and producing poor-quality material.

The effectiveness of such "in process" inspection is illustrated by a manufacturer of aluminum die cast parts, which occasionally produced a few bad castings. In some cases the castings were dimensionally incorrect because of warping. In others, the castings had porous areas or were not completely filled in, leading to surface problems. Prior to SPC, the operator of the machine would take each casting as it came out of the machine, remove excess metal, and toss it into a metal tub where it was later inspected by quality-control people. When SPC was put in place, the operator was given a check fixture that identifies dimensional problems. The operator now removes excess metal from each part, inserts the part in the check fixture to be sure it is dimensionally correct, visually checks the surface for problems, and approves only the good parts. Rejects are immediately melted down so they won't be shipped to the buyer. This is the goal you are trying to achieve through SPC.

Finally, as introduced in Chapter 2, you will want the supplier to chart the quality results of each lot of material produced. This will alert him to potential problems before the process gets out of control. Figure 6.1 demonstrates this technique.

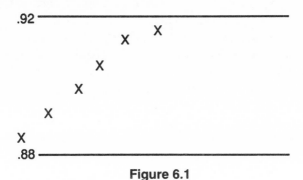

Figure 6.1

The acceptable tolerance on a certain element of the specification is .88 to .92. As long as the supplier produces product within this range, it is acceptable. The x's represent the results of each lot of material produced. Note that over the six lots produced to date, the trend is pointing out a potential problem. The process is going out of control, the tool is not sharp enough, something is allowing the quality of the part to get near the outer range. By charting the process, the supplier should be able to see this trend and correct the process to get future lots back in the center of the range. (See Figure 6.2.) This will help assure you of quality parts lot-to-lot from the supplier.

Once the supplier is charting all the key elements of the specifications and has the process under control so it is only producing good material, the supplier should begin sending copies of the charts to your Quality

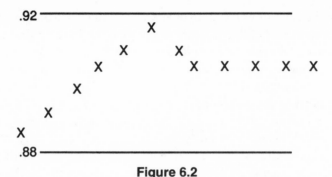

Figure 6.2

Assurance (QA) Department. Your QA people should test those same elements and develop a confidence that the supplier has performed the tests correctly. After confidence has been developed in the supplier's charts, reduce incoming inspections and only randomly check the material for a given period of time. If no defects are found, certify the quality of that material for that supplier and eliminate incoming inspections (you will also eliminate the indirect costs that do not add value to the product).

Note that only the specific material is certified with the supplier. Don't assume just because the supplier can make one item with no defects that he can also make all other items defect-free.

In most JIT/TQC companies that accept control charts, periodic checks are done at the supplier's facility to verify that the quality-control tests are being run in an acceptable manner. It is a good idea to have your quality-assurance manager and your supplier's quality-assurance manager know each other personally. Then, if there are any questions on raw material, processes, testing, etc., the two QA people can quickly discuss the problems before the supplier makes an unacceptable quality item.

Finally, if the supplier is going to put through an engineering change on your item, your QA people should check the change before it becomes effective. If that change causes you a problem in fit, function, or use, the supplier should not make the change. It is better that you know the effects ahead of time rather than when it is too late, when your product fails in the field because of the unknown change.

THE PROCESS WORKS

Tektronix had an ongoing series of quality problems with its suppliers. As a result, it employed one hundred people in supplier quality control and acquired a good deal of very expensive test equipment to check incoming material. Although this company had fifteen manufacturing plants, all purchase items were brought in through one central receiving/ inspection area. After being checked for quality and quantity, the materials were stored at a central receiving area in a very large computerized warehouse and then shipped to the individual plants as required. Not only did this company have a very large Quality Control Department at the central location, but it needed more than one hundred people in Receiving and Warehousing to move and store the items.

To improve the situation, the company formed a multidisciplinary team that included Quality Control, Receiving, Warehousing, Industrial and Product Engineering, Tooling Engineering, Finance, and Marketing. Purchasing chaired the team. The team's first step in addressing the quality problems was a series of "kickoff" education days for the suppliers' top management to explain the issues and the new TQC approach to quality. The ultimate goal, they said, was defect-free product delivered frequently to point-of-use.

Once the suppliers' top management agreed to the program, a second series of kickoff meetings was held for the suppliers' operating managers. The team then started visiting each supplier to check the manufacturing processes and tooling and identify deficiencies that lead to defects. As a defect was identified over the next several months, it was given to the supplier for cause identification and elimination. As the quality levels improved at the suppliers, the expensive test equipment was transferred to the suppliers and the supplier was required to verify the quality and deliver only defect-free product. As the suppliers developed SPC and control charts, the company shifted to random inspections and, after a four-year process, certified the quality on the majority of their purchased items.

Today this company employs one incoming inspector at each of its fifteen manufacturing plants and five source quality inspectors who visit supplier plants randomly. All certified items are delivered directly to point-of-use on the manufacturing floor as required, without going through Receiving or Inspection. The large central warehouse is currently for sale because the receiving, inspection, and warehousing functions are no longer needed on the majority of the items. The few items that aren't yet certified are delivered, as required, to the incoming inspectors at the plants. The goal is to eventually have 100 percent of the items certified.

As each supplier becomes certified for producing defect-free product, they are given a plaque to celebrate the occasion. This plaque is a symbol of an ongoing commitment by both companies to the highest levels of quality and performance.

SUMMARY

- In companies that consistently waive and use defective material, the suppliers are given the message that it is okay to ship defective material.

- Supplier certification entails a three-step process that consists of qualifying and educating suppliers prior to the certification step.

- Without management support, a supplier certification program will not be successful.

- To be effective, a cross-functional team should be used to qualify suppliers.

- To be able to deliver defect-free material, suppliers will have to begin using quality techniques like statistical process control.

- Control charts are a good way to visually alert suppliers to potential quality problems.

- If your supplier is going to make an engineering change on your item, he should check with you first to verify that the change will not cause a problem in your product.

Chapter Seven

Supplier Base Reduction

Many companies today have a policy that states that Purchasing must have more than one supplier approved on each item purchased. The logic underlying this policy is simple enough. First, if you have more than one supplier, the price of the item can be regularly bid so that the company is assured that it is getting the lowest price for the item. Second, if the supplier has a production problem, the alternative supplier can step in and keep the company running.

In principle, that reasoning is sound. But it doesn't work in practice. The first reason mentioned for multiple suppliers is to get the lowest price. The idea of competition is good and we should always be looking for the "best" supplier. But that doesn't necessarily mean the supplier with the lowest price; you will never get to lowest cost by a quotation form. You get to lowest cost by working in a partnership with your suppliers to eliminate all waste and thereby developing the lowest cost *consistent with required delivery and quality requirements.*

Second of all, having multiple suppliers doesn't prevent stockouts. Regardless of how many suppliers they have per item, companies still have shortage lists and run out of purchase parts on a regular basis. Why? Because they don't have the best suppliers, they don't train those suppliers in what is expected performance-wise, and they don't have valid schedules. It is difficult enough to find one good supplier who can consistently perform, let alone two or three suppliers. When purchasing people are forced by company policy to find multiple sources for all items, they usually have to dip into the pool of poor or marginal suppliers to make their quotas. This only guarantees that some of those suppliers will not perform up to expected performance levels. Also, it is impossible to adequately educate and train several hundred or several thousand suppliers. With a couple hundred suppliers, that becomes a more manageable task. One company's motto is "Make the suppliers so good they can't lose the business." Of course, that can't happen with a large supplier base.

How much can you reduce your supplier base? Consider Xerox, which cut back from 5,000 to 300 suppliers over a couple of years' time span. Then it invested heavily in training those 300 suppliers in MRP II and JIT/TQC. The results? Xerox cut its quality defect level from 10,000 (1 percent) to 122 (.001 percent) parts per million, cut its raw material inventory 75 percent, dramatically improved its on-time supplier performance, was able to link up with its key suppliers using EDI, and reduced the purchase content of a copy machine 48 percent. Not bad performance numbers! Tellabs, a manufacturer of high-speed telecommunications products based in Lisle, Illinois, cut its supplier base from 600 to 150 and reduced the number of trucking companies that carried the raw materials in half. With fewer suppliers to manage, the company can now give its MRP output directly to the suppliers and have them do the purchase material planning. The direct labor employees release purchase material as needed from those suppliers using Faxban. Purchasing now buys material by exception only, when there is a problem, and has time to work very closely with those suppliers on how to continuously eliminate waste and lower the total costs of acquisition.

Finally, Unisys, of New Jersey, cut its supplier base from 750 to 106 and reduced the number of trucking companies from 120 to 8. As a result, purchase costs have dropped by 40 percent, freight costs have been reduced from $.32 per pound to $.09 per pound, and transit times from the Far East have been cut by two-thirds.*

If done correctly, reducing your supplier base can have dramatic results in purchasing.

Single-Sourcing Versus Multiple-Sourcing

There are many excellent reasons why single-sourcing is more beneficial than multiple-sourcing. We will cover only seven advantages in this section, but every company will be able to add some to the list due to its own particular circumstances and supplier base.

One of the first advantages is consolidated volumes. If you split the

* *Purchasing Week*, July 16, 1987.

business among three suppliers, the pricing typically will not be as good as buying the entire volume from one supplier. Whether you give each supplier 33 percent of the business or split the business 70 percent to Supplier 1, 20 percent to Supplier 2, and 10 percent to Supplier 3, the price will not be as good as if one supplier receives the entire volume. Also you run the costs up because you will need to negotiate more business agreements and have more receipts and quality inspections. You probably will have a higher freight rate because individual shipping quantities from Suppliers 2 and 3 are smaller, too. In addition, the costs of training and certifying multiple suppliers will be higher, and many JIT techniques such as point-of-delivery will be almost impossible to implement.

Also, it is often very difficult to find several good suppliers for an item. Many surveys have been published that show that the typical company experiences supplier performances that range from 70 to 85 percent on-time and 87 to 93 percent in terms of in-specification quality parts. A buyer in a supplier scheduling environment cannot tolerate that level of supplier performance because the supplier schedule or Kanban mechanism is working to the need date of the item, not a replenishment date. Either the supplier hits the date with acceptable parts or the buyer has a potential stock-out situation. As the buyer is forced by company policy to buy from two or three suppliers, the chances for good supplier performance drops.

Next, as mentioned before, buyers need to either train suppliers or ensure that they get training in MRP II and JIT/TQC. Many suppliers have little or no concept of how to produce defect-free items, make consistent on-time deliveries, or reduce lead times or lot sizes economically. Therefore, if the buyer wants better supplier performance, training must occur. (See Chapter 13.) If a company has a thousand suppliers and educates twenty at a time, that tallies up to fifty days or ten solid weeks' worth of training effort. If a company reduces the supplier base to two hundred, the training load drops to ten days or two weeks, a much more manageable work load.

In Chapters 4 and 8, we cover the need to move toward more frequent shipments and the concept of synchronizing your schedules with those of your suppliers. If you can get suppliers to set aside capacity to support your supplier schedules and then allow the buyer to select out of that capacity the material needed on a weekly or daily basis, then your company can become much more flexible at meeting its customer needs without carrying a lot of "just-in-case" inventory. The more suppliers a buyer has, the smaller the volume of parts per supplier, and the less sure the supplier can be about long-term guarantee of business. The latter means it will be more difficult to implement synchronized schedules.

Because there is only one supplier for an item, the communications between the buyer and the supplier are often much easier. Typically, with a large supplier base, the only person the buyer knows from the supplier is the salesperson. If a problem occurs, the buyer calls the salesperson to discuss the problem. The salesperson calls Customer Service, which calls Quality Assurance, which calls someone else. This is time-consuming, frustrating, and seems to take forever before an answer is provided. With only a hundred and fifty suppliers, it may be possible to get to know the key people in Order Entry, Quality Assurance, etc., at the supplier on a first-name basis. The problem can then be given directly to the correct person and resolved in a much more timely and efficient manner.

Also, a smaller supplier base allows more of a joint approach to problem-solving. The buyer and supplier are able to work more closely and, in effect, see themselves as a team fighting against the competition. The better the two perform together, the more business they are jointly able to generate. There is a joint commitment to get the order, to build a quality product on time that will get repeat orders, and to jointly make a profit off the order.

When there is a single source for an item, it is easier to get the suppliers involved earlier in the product-development process, which can result in significant improvements in quality, cost, and delivery. In addition, with fewer suppliers involved, it is easier to get your suppliers involved in problem-solving sessions. One Midwest company had the opportunity to bid on a one-time order for 125,000 units of a product it produced on a regular basis for other customers. The company normally made 500,000 of the units annually, so if it could get the order, there would be a 25 percent increase in volume for all the suppliers as well. The best price the company could come up with and still make a small profit on the order was around a hundred dollars per unit, but at this price it probably would not get the order. To try and determine how to offer a more competitive price on this one-time order, they held a meeting that was attended by all the key internal managers as well as all the key supplier personnel. They jointly discussed what could be done, on a one-time basis, to reduce the cost of the product both internally and externally. Approximately twenty dollars in cost was identified and, in a team effort, removed from the product. The unit was bid at ninety dollars each, the contract won, and everyone was able to share in added business and profit that would not have been realized otherwise.

Finally, the buyer is now able to purchase by exception only. With fewer suppliers, all working either to supplier schedules or Kanban, the buyer is removed from the daily routine of paperwork and expediting and only needs to get involved in the day-to-day routine when something goes wrong. The buyer now has time to spend developing and educating suppliers, negotiating long-term contracts, working with suppliers to simplify their products and processes, and integrating the suppliers into the business. If you can develop a partnership with a few good suppliers who deliver consistently on-time and in-specification with few problems, you finally can get the time to drain the swamp and uncover new opportunities to improve performance in the future.

Single- Versus Sole-Sourcing

Sole-sourcing indicates that only one supplier is approved to produce a part. Single-sourcing indicates that more than one supplier may be approved to produce a part, but only one supplier is *used* to produce the part. We are not suggesting in this discussion that you sole-source all your parts, but rather that you single-source.

Consider a multi-sourcing environment that involved nine suppliers. (See below.)

In this example, you have three approved sources and purchased, where practical, product from each source during the year on each item.

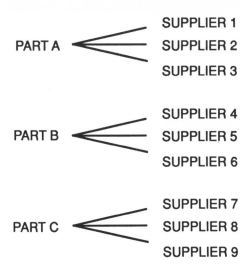

PART A — SUPPLIER 1 / SUPPLIER 2 / SUPPLIER 3

PART B — SUPPLIER 4 / SUPPLIER 5 / SUPPLIER 6

PART C — SUPPLIER 7 / SUPPLIER 8 / SUPPLIER 9

Parts A, B, and C were all the same commodity, plastic injection molded parts, but differed in either size, color, type of plastic, etc. Supplier 1 was your best supplier in terms of overall performance, followed fairly closely by Supplier 5, with the other suppliers significantly lower in performance. In a single-source environment, then, you would purchase all parts from Supplier 1.

But note that you have two choices at this point. You could consider Suppliers 2 through 9 as approved suppliers should a problem arise, or take Supplier 5, the second best performance supplier, and approve him on Parts A and C. That way you would not be sole-sourced, but we would be buying from a single supplier for all the reasons listed earlier in this chapter. The advantage for Supplier 5 in this situation is the potential for future business on other than Part B, and the opportunity for developing expertise using the training given by the buyer. In short, this becomes a win-win situation for all involved and eliminates a lot of time the buyer now spends trying to manage too large a supplier base.

Operating Under Single-Sourcing

What is it like operating under MRP II and JIT/TQC with single-sourcing? One buyer indicated that she would never work for another company that didn't have it in place. When asked why, the answer was, "Quality of life." When you have valid schedules, a manageable supplier base, predictable performance, and time to do your job, work is certainly more enjoyable and rewarding.

One issue needs to be addressed at this point: pricing and profit protection for the supplier under a single-source arrangement. The concern is that the supplier, knowing he is the single source, will try to raise the price and make an excessive profit. On the other hand, the buyer, having been in the supplier's operation and knowing a lot about the supplier's costs, will try to drive the price down so the supplier makes an insufficient profit. The supplier deserves a reasonable profit and the buyer deserves a reasonable price.

With this in mind, there are three alternatives in pricing. The list below shows the supplier's cost breakdown that will be used in the three alternative examples:

PART A—Price $1.00 each

Cost Breakdown

Raw Materials   .40 each   (40%)
Direct Labor
   and Burden   .40 each   (40%)
Supplier Profit   .20 each   (20%)

The first alternative says that the buyer demands a price reduction from the supplier and gives the need for this price reduction, but does nothing to help the supplier reduce the costs. To retain the business, the supplier reduces the price to $.90 each. Because the supplier has the same costs for raw materials and direct labor as before, the reduction must come out of the supplier's profit. The profit is cut in half, from $.20 each (20 percent) to $.10 each (10 percent). The supplier may try to cut corners to lower costs, but that may end up causing the buyer quality or delivery problems in the long run. Because the buyer's business has just become less attractive to the supplier, service, performance, and long-term relationships will suffer as a result.

The second alternative says the buyer works with the supplier to reduce costs on the item, but keeps the percentage of profit fixed on the part. Working as a team, the buyer and supplier are able to reduce both labor and raw materials by 25 percent. The pricing would now be:

PART A—Price $.75 each

Cost Breakdown

Raw Materials   .30 each
Direct Labor
   and Burden   .30 each
Profit (20%)   .15 each

While the percent of profit remains the same, the actual dollars of profit are reduced for the supplier. While the supplier is still profitable and better off than alternative one, the supplier will not be aggressive in reducing the costs because it reduces the total dollars of profit.

The third alternative says that the buyer and supplier work together as a team to reduce the cost of the item, but the amount of profit is fixed on the item. Like the second alternative, the labor and raw materials is reduced 25 percent, but the supplier is guaranteed a $.20 each profit on the item. The pricing would now be:

PART A—Price $.80 each

Cost Breakdown

Raw Materials   .30 each
Direct Labor
    and Burden   .30 each
Profit (Fixed)   .20 each (25%)

Note that although the buyer's price is slightly higher, the supplier's profit percentage is up and the total dollars of profit on the business is unchanged. The advantage to the supplier: guaranteed profit on the business and potentially more volume in business because the buyer's company can be more competitive in the marketplace. The advantage to the buyer: a 100 percent pass-through on all cost reductions to the buyer, a supplier who has no incentive to not reveal the cost breakdown on items, and a supplier who now has an incentive to work cooperatively with the buyer to reduce any and all costs that do not add value to the product. This is truly a win-win situation in which the buyer is able to secure lower costs with the supplier's complete support of the effort while the supplier is able to maintain a level of profit.

An East Coast buyer had a major problem with this concept. The example quoted was one in which, over four years, the raw material cost and direct labor costs were reduced by half. The buyer's objection was that the supplier was now making a 33 percent profit, and that was unacceptable to the buyer. The item quoted consisted of the following costs:

PART A—Price $.60 each

Cost Breakdown

Raw Materials   .20 each
Direct Labor
    and Burden   .20 each
Profit (Fixed)   .20 each (33%)

The buyer stated he would never allow a supplier to make that kind of profit on an item and felt the supplier's profit was the supplier's problem, not his as a buyer. When questioned further, this buyer admitted that he was proud of the fact that he often switched from supplier to supplier and had no need to develop partnership relationships. Upon further questioning, the buyer conceded that he had a fair number of past due orders, quality problems, and had only generated about a 2 percent cost reduction for the year, which was less than the goal he'd set for himself. That is typical of an adversarial relationship with the suppliers. The win-win partnership relationship yields nearly 100 percent on-time, in-specification parts, and a 13 percent annual cost reduction. The results speak for themselves.

Pick a group of like items, single-source those items, educate the suppliers in what you are attempting to do, generate valid schedules for that supplier, approach pricing in the win-win approach described above, and you will be amazed by the results.

SUMMARY

- You will never get the lowest cost by a quotation form.

- There are several reasons why single-sourcing is advantageous over multiple-sourcing.

- With fewer suppliers, it is easier to get your suppliers involved in education, problem solving, synchronized schedules, and product development.

- Single-sourcing is preferable over sole-sourcing.

- If supplier partnerships with single-sourcing is going to work, the supplier must receive a reasonable profit and the buyer a reasonable price.

Chapter Eight
Synchronized Deliveries

In earlier chapters, all the Kanban examples covered daily deliveries from suppliers. That is practical when the supplier is fairly close to the buyer's company and the travel time is short. When the supplier is further away, though, daily shipments may not be practical and cost-effective. Also, your company may not need daily shipments on all the items you buy. The objective is not to secure daily shipments, but frequent shipments that are synchronized to your production schedule in a cost-effective mode.

At some companies, for example, a box of a thousand screws might be a week's worth, so it would be overkill to have them delivered daily. On the other hand, consider a West Coast manufacturer of high-volume, expensive test equipment that consumes thousands of fasteners a day. This company has thirty different configurations of fasteners that are used at several locations on the shop floor. The cost of receiving, warehousing, issuing, and keeping track of all these fasteners was very high. Therefore, the company sought to develop a daily delivery of all the fasteners, have the fasteners delivered to the point-of-use on the shop floor, and eliminate the need to receive and warehouse the fasteners.

The company now uses a technique called the "bread run." By analogy, consider how a bread truck arrives every morning at your local supermarket and replenishes the loaves on the shelf. The driver then reports to the manager how many loaves it took to restock the store. The West Coast manufacturer just referred to uses the same mechanism to ensure an adequate supply of fasteners, setting up Kanbans—small bins—at all the manufacturing floor locations where fasteners are used. It then developed a plant map to show which fasteners are used, and where they are used. Each day, the fastener supplier arrives, follows a predetermined path around the plant, fills up the bins at every location with the required fasteners, and then gives Receiving the counts he delivered.

This process has eliminated receiving, warehousing, parts movements, and paperwork in favor of daily deliveries (which may or may not be appropriate for every item you use). The company then does a daily comparison in their MRP II system between what was used and what the supplier reported he delivered to each location. If the two numbers are within an acceptable tolerance, the Accounting Department pays for the delivery.

The same manufacturer buys one of its key components from Japan. Each part costs several hundred dollars, and the company uses about two hundred of them per week. The company used to import the parts through the West Coast but had many problems with customs clearance. The solution? A lot of inventory on the part, just in case additional shipments became snarled in red tape. After spending a lot of time and effort, the company was able to locate and work with the customs officials at an East Coast port to smoothly and quickly move the parts through customs on a weekly basis. The customs officials have become used to seeing the parts arrive, and quickly release the shipments as soon as they hit the port. The parts are then moved from customs to the local East Coast sales office of the Japanese manufacturer.

Each Friday afternoon, the West Coast buyer for the parts faxes a release to the East Coast branch of the supplier, detailing how many of the parts are needed for the next week's production schedule. The West Coast buyer has established a contract with a trucking company to pick up the parts late Friday afternoon on the East Coast and deliver them to the point-of-use on the manufacturing floor in California at 8:00 Monday morning. Weekly the parts are brought in for that week, they always arrive at 8:00 A.M., they are always delivered to the point-of-use on the shop floor, and the just-in-case inventory has been eliminated.

With this arrangement, the freight costs are actually lower than the total costs associated with the old method of shipping the parts from Japan to the West Coast plus the cost of the just-in-case inventory. Most important, the company now has a guaranteed weekly receipt quantity that is delivered directly to the point-of-use on the shop floor at 8:00 each Monday morning.

Shifting to daily deliveries in this case would not be necessary, nor would it be practical. But note that shipments were *synchronized* with the buyer's schedule (the amount needed for the week was delivered at the start of the week). As a result, they were more cost-effective (total cost now lower per delivery), and they were more frequent (weekly versus monthly).

Every company needs to decide on the frequency with which it wants—and needs—deliveries from suppliers. In general, though, most companies need to move toward more frequent shipments from their suppliers in order to reduce the costs ("waste" in JIT terminology) associated with the inventory investment it has tied up in purchase material.

FREIGHT CONSIDERATIONS

One of the major drawbacks to more frequent shipments can be the cost of transportation on the purchase material. If the transportation area is not managed correctly, freight costs can skyrocket and negate any benefits in purchase price reductions and inventory reductions obtained by adopting MRP II and JIT.

Let's look at a situation that illustrates this point. (See Figure 8.1.) A company based in Grand Rapids, Michigan, purchases forgings from a supplier in Detroit, a three-hour drive away. The forgings are purchased in a monthly lot quantity of 5,000. Each of these forgings weighs four pounds, so the monthly freight quantity, per shipment, is 20,000 pounds. The truckload rate for this shipment was $1.96 per 100 weight, or $392.00 for the total shipment.

Purchasing, in its initial JIT effort, cut the lot size to a weekly quantity, or 1,250 per week. It was able to negotiate the same price for the weekly shipment as for the monthly shipment, so the new approach seemed wise. But the weekly shipment weight dropped to 5,000 pounds, and the freight rate on that quantity increased to $4.43 per 100 weight, a 126 percent increase. In Purchasing's continued effort to reduce the lot size, it asked for and received daily shipments at the same price from the supplier. Again, everything appeared on target, except that the daily shipment weight was 1,000 pounds. The freight rate on 1,000 pounds was $8.21 per 100 weight, a hefty 319 percent increase over the truck-load rate. At the end of the month, the freight bill for the daily shipments totaled $1,642.20, which negated any inventory savings the company was able to obtain by going from monthly to daily shipments.

Forgings Freight Costs			
Shipment Frequency	Weight per Shipment	Cost per 100 Weight	Monthly Cost
Monthly (1 per mo.)	20,000	1.96	392.00
Weekly (4 per mo.)	5,000	4.43	886.00
Daily (20 per mo.)	1,000	8.21	1642.20

Figure 8.1

As you can see, the tradeoff can be problematic if transportation costs are not managed well. Most companies today will specify which common carrier they want to deliver the items and may negotiate the freight rate, but few go much further than that. Here are some JIT techniques being employed in the freight and packaging areas today.

Pooled freight rates is one technique that can be employed to lower freight costs. In this technique, we take all the shipments from a given area of the country, combine them into one shipment at a consolidation point, then negotiate one freight rate based on the total tonnage of that load. For instance, a Midwest-based company had several small suppliers in the Chicago area that had begun to ship them weekly with less-than-truckload shipments. Each shipment ranged in weight from around twenty pounds to more than four thousand pounds, depending upon the supplier and part number involved. Because the company was getting weekly shipments rather than, say, monthly quantities, the average weight per shipment was less so that higher freight rates per hundred weight applied.

To remedy the problem, the buyer selected one common carrier based in the Chicago area that ran regular shipments into that part of the country. The buyer was then able to negotiate a pooled freight rate based on the total weight of the combined weekly shipments from the Chicago area to their location. All the suppliers were then instructed to arrange for that common carrier to pick up their shipments on Tuesday, and all of the small individual shipments were consolidated into one large shipment, which was delivered Wednesday morning to the buyer's plant. Because the total tonnage is now greater than the individual smaller shipments they had before, the buyer was able to get smaller, weekly shipments but pay less total freight costs.

Another variation on this technique is sometimes referred to as the "milk run." Instead of the consolidation occurring at the common carrier's location, the company hires the common carrier on a contract basis to carry out the consolidation. A Denver company, for example, has several suppliers located in small towns two or three hundred miles from Denver. This company contracted with a freight company to stop at all those locations once a week and pick up the purchase parts. Monday morning the truck makes a pickup in Town A, Monday afternoon in Town B, and so on. The trucker stops at ten companies during a three-day period and delivers the material back to the buyer's plant late Wednesday evening. The company pays the trucker a flat fee for the run, regardless of whether the truck comes back full or not. In effect, it rents the truck for the three days. The total freight costs to rent the truck for the three days are dramatically less than what they used to pay for the ten individual shipments, and this allows the company to bring in what it needs to support the weekly production schedule without paying more in freight for the more frequent service.

Companies that own their own trucks would ideally like to backhaul purchase parts so the trucks do not come back to the plant empty, but coordinating this backhaul scheduling is often difficult and not as productive as many companies would like. With supplier scheduling, though, the coordination is much easier. The supplier schedule is generated on a weekly basis for all their suppliers. The quantities due from the suppliers each week are multiplied by the piece weight of each part, and an inbound freight schedule is developed. This can be summarized by town code, county code, state code, etc. Outbound trucks are then routed in a manner that achieves maximized backhaul potential. Those shipments that will not be picked up by your own truck are shipped common carrier. In essence, the backhaul is free, so the freight costs are totally eliminated.

One Midwest company devised a very creative arrangement in this area. This firm had many small suppliers on the East Coast. It also had a large dealer for its product in the Boston area to which it delivered product on a weekly basis with its own truck. Purchasing rented two thousand square feet of warehouse space from that dealer and instructed all of its Northeast suppliers to ship to this dealer on a weekly basis. The parts had to arrive by each Wednesday, because on Thursdays the company truck delivered product to the dealer and backhauled all the purchase parts. Consequently, the company was able to obtain weekly shipments from all its Northeast suppliers and yet reduce freight costs by more than 80 percent over the old, less-than-truckload way of doing business.

Another twist on the same theme occurred with a buyer in Minneapolis. His company had a supplier in central Wisconsin that shipped a truckload of purchased material each week. To reduce freight costs, the buyer attended several different purchasing association meetings in central Wisconsin to determine whether any companies had shipments coming out of the Minneapolis area. Fortunately, the buyer located a company that purchased a truckload of material a week from Minneapolis, and the two firms contracted with a trucking firm to do one round trip weekly, carrying the material each way for the two companies. Because the trucker was guaranteed a full backhaul for the trucks, they were able to negotiate a lower truckload freight rate. In the final analysis, both companies were able to lower their freight costs by 30 percent.

One other technique is being employed by JIT companies, whether common carriers or contract haulers are used. This is point-of-use delivery, rather than delivery to the dock. Most companies today have parts delivered to the dock, where they're counted, inspected, moved to the warehouse, and then later moved to the manufacturing floor. All of those operations add cost to the process, but no value. Ideally, in a JIT environment you can eliminate all non-value-adding processes. Where companies have certified the quality of their suppliers so no inspections are necessary, parts can be delivered directly to the point-of-use on the factory floor, thereby eliminating those non-value-adding operations. (See Figure 8.2.) In effect, all of their certified suppliers do a "bread run" type of delivery as discussed earlier in this chapter.

How can you get point-of-use delivery if you use a common carrier?

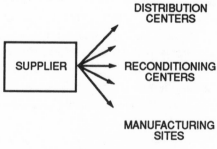

Figure 8.2

Companies specify the driver they want, and then train that driver as to where the parts are to be delivered on the shop floor. To use this technique with a common carrier, negotiate a JIT contract stating that parts must be delivered to specific locations on the shop floor, and that John Doe is the only driver allowed under the terms of the contract. What happens if John Doe is sick or on vacation? Either train an alternate driver or specify that the parts must be delivered to the dock if John isn't available for that delivery.

Will more frequent shipments work if you import materials from overseas? Yes, but it will typically take a lot more time and effort to put the arrangements together. One East Coast manufacturer of computers purchases 85 percent of its components from Asian suppliers. The company selected a customs broker located in the Far East to coordinate the shipments. Here's how it works. Monday of each week, the buyer sends a Faxban to each of the Far East suppliers, detailing what is to be shipped that week. These suppliers then forward their components to the customs broker, who consolidates each of the individual shipments into a single container shipment. The shipment leaves the broker's location at the end of the week and arrives the next week at the buyer's plant. To do all this, the buyer has had to coordinate each shipment with the suppliers, customs broker, customs officials, the air freight forwarder, and the local trucking company. Is all the effort worth it? This company has been able to lower its incoming freight costs by over 40 percent!

The reverse situation also works. An Australian manufacturer of water heaters purchases virtually all of its components from suppliers in the United States. The company hired a firm in Los Angeles to consolidate the components from all of the suppliers into a single container and forward them to Australia each week.

Whatever your shipping needs, the basic message is be creative, challenge the way you're doing it today, and look for better ways to manage your freight costs.

PACKAGING AND LABELING

Very few companies take advantage of the benefits possible in this area by standardizing carton sizes and labeling requirements. They allow the supplier to package product in any size container and to vary quantities from container to container. The buying company asks the supplier to label the container with the part number and quantity, but it does not specify the location or type of label required. As a result, this company misses out on ways to reduce costs through packaging and labeling practices. This is illustrated by the following example.

A motorcycle manufacturer produces motorcycles in lots of one hundred. Every supplier ships in cartons that contain one hundred parts. The company issues one container of each part to the assembly area, and at the end of the manufacturing run, no parts are left over. The stockroom doesn't have to count parts into or out of the stockroom; it issues and receives in cartons of one hundred, which makes its job a lot easier.

This system also makes receiving parts easier. Receiving counts the number of cartons and multiplies by one hundred to ascertain the receipt quantity. How does it verify that the supplier counts are really one hundred per carton? At the end of the manufacturing run, if there are none left on the manufacturing floor and one hundred motorcycles were produced, there had to be one hundred in each carton. Wherever possible, standardize carton sizes and carton quantities around your manufacturing lot sizes. This will eliminate a lot of wasted time and motion in receiving and counting the material.

As you move to JIT deliveries, the workload for Receiving will increase as you get more frequent shipments. You therefore need to develop JIT approaches that reduce workload, not increase it. To reduce the workload and make it easier for Receiving, companies have gone to bar coding. The buyer provides the supplier with bar-code labels that define the supplier number, item number, and quantity in each container. (See Figure 8.3.)

Figure 8.3

The supplier puts one bar-code label on each container. When the shipment arrives at Receiving, the Receiving person scans the bar code label on one container and records the number of containers. At this point, receipt is complete. This results in more frequent shipments but less total effort for Receiving. Also the bar-code label facilitates point-of-use deliveries. The parts are delivered directly to the shop floor and the operator in that department can report the receipt directly to the computer using a bar-code reader. In deliveries made directly to the point-of-use, Receiving doesn't even need to be involved, so the cost of receiving is basically eliminated from the process.

Again, the goal of JIT is the identification and elimination of waste in all functional areas of the company. That means constantly attacking waste, making lots of little changes, and constantly improving procedures until there is no waste of any kind in the system. That also means looking for waste in the following areas: order quantity, safety stock, rejects, setup, receiving of parts, movement of parts to warehouse, issuing of parts to floor inventory, and, as we have seen in this chapter, transportation.

Summary

- The objective is not necessarily to secure daily shipments from suppliers, but frequent shipments that are synchronized to your production schedule in a cost-effective mode.

- If not managed correctly, freight costs can skyrocket and negate any benefits in purchase price reductions and inventory reductions obtained by adopting MRP II and JIT.

- There are several techniques, such as bread runs, milk runs, and pooled freight rates, that allow the buyer to get more frequent shipments at a lower freight cost.

- Where companies have certified the quality of their suppliers so no inspections are needed, parts can be delivered to the point-of-use on the shop floor.

- By standardizing packaging and labeling, the costs of receiving can be reduced even though you are receiving more frequent shipments.

Chapter Nine

The Customer/Supplier Relationship

Prior to developing partnership arrangements, buyers and suppliers often have adversarial relationships. One reason is that the buyer's schedules are not valid and as a result are constantly changing. The buyer knows that the dates on the purchase orders are replenishment dates, not need dates. The buyer therefore knows that even though an order might go one or two weeks past due, the company will not run out of material. The flip side of the coin is that the supplier knows the dates aren't valid. This is obvious to the supplier because no one is screaming for the past-due orders. Eventually, the buyer's company runs out of material, the item hits the shortage list, and the buyer aggressively expedites the past-due order. And who takes it on the chin for missing the invalid date? The supplier.

Another source of the adversarial relationship is that the buyer's quality specifications are not always representative of what is really needed. Let's say that a supplier shipment is initially rejected because it is outside certain tolerances, but since the parts are badly needed, the rejection is waived and the parts are used. At the time of the next production run, the supplier experiences the same quality problem. The supplier ships the parts to the buyer knowing they are out of specification, and they are again rejected. This time the buyer has a little time before the inventory of the part is depleted, so the buyer wants to return the parts to the supplier. The supplier then wants to know why the company could use the parts from the previous shipment, but not from this one. The relationship deteriorates a bit further.

At this point the buyer requests that the supplier consider lowering the price on the item purchased because of some vaguely defined competitive pressures. The supplier is not assured of any benefits by lowering the price and begins to question the soundness of a demand that only benefits the buyer. But the supplier also realizes that the buyer has several other sources for this item and could easily give some other firm the business. This transmits the message that the buyer has no real loyalty to the supplier, even if the supplier lowers the price. Both parties thus see each other as the aggressor, and the adversarial relationship becomes the standard way of doing customer/supplier business.

In a partnership relationship, there is a formal plan to which both parties have agreed. The dates on the supplier schedule are valid need dates. Moreover, the quality specifications define the minimum requirements and do not have unneeded or unrealistic tolerances. Common objectives form a win-win relationship. The supplier is assured of the business and a fair price for as long as the partnership lasts. The buyer is assured of on-time delivery of defect-free parts. A communication network ensures that information flows freely between the buyer and supplier.

Also in a partnership arrangement, the supplier schedule is frequently communicated to the supplier and time fences are respected. The supplier regularly communicates any problems or exceptions to the plan as they occur. Performance measurements that have been jointly agreed to are used to track how well both parties are doing. The measurements are not used for punishment or reward, but rather for planning performance improvements.

Partnership arrangements are characterized by a sense of teamwork, cooperation, and a shared goal of continuous improvement. This leads to more predictable performance, because all of the problems that have occurred in the past have been identified and eliminated. Finally, partnership arrangements improve the quality of life in the workplace by eliminating the adversarial basis of doing business.

GUIDELINES FOR EFFECTIVE PARTNERSHIPS

The following guidelines should serve as the backdrop for all of your supplier scheduling efforts. If you observe them in all of your supplier scheduling actions and plans, you will maximize your chances for a high return on your investment of time and energy.

1. *Trust and communication has to be open and honest.* This is by far the most important of all the rules. If the buying company or its suppliers have their own "hidden agendas," you can be sure that the entire system will at some point fail.

2. *Suppliers are part of your organization—they just happen to be outside the gate.* The traditional organizational chart ensconces itself in a moat. For supplier scheduling to work, the moat must be bridged and suppliers brought back into the fortress. Only when suppliers are viewed as critical players can a scheduling system work.

3. *Shared risk means a fair price and a fair profit.* The traditional "zero-sum game" must stop. You can't always get the lowest prices and suppliers can't always seek the maximum margins. Everyone comes out a winner when they make concessions in favor of a long-term relationship. (See next point.)

4. *Long-term relationships are better than one-night stands.* When a company and its suppliers take their vows, they ensure themselves of sustainable profits and benefits—predictable quality, predictable cash flow, and predictable growth and development.

5. *Partnerships consist of a great deal more than signing papers or issuing some edicts.* A partnership between a company and a supplier is truly a culture change, a different way of running the business. But you can't just mandate a new culture and expect it to materialize. People must be educated about changes in the way they do their jobs. Systems must be in place. In short, management must regard itself as an agent of change, not just the authority that dictates change.

6. *You can't mass-produce partnerships.* Supplier scheduling won't work with all suppliers. You need to pick them one at a time and see which ones work and which don't. And don't start off with a written contract that may lock you into a relationship you may soon regret. Begin with a handshake and ease your way into a signature.

7. *Performance is the main criterion for picking a supplier.* Size is often irrelevant to whether or not a supplier can deliver the right goods to the right place at the right time. Look at performance first; everything else is a footnote.

8. *Internal accountability is a key factor.* When you develop a supplier scheduling system, it's critical to establish who's responsible for what. The system demands that everyone has an allotted role and knows what it is.

9. *Sell the concept to management, get the necessary leadership and support.* A supplier scheduling system will greatly reduce the number of suppliers your company deals with. This is a major change and requires top management support for it to be successful.

10. *Use an existing supplier to kick off the system.* Evaluate your current suppliers and pick the most likely candidate for a partnership arrangement. Sit down and negotiate the terms, conditions, and price; i.e., include everything you need to spell out in the partnership.

11. *Develop the working relationship.* All successful business relationships are based on ground rules. How do we communicate? How do we solve problems? These and other issues must be resolved on both sides of the partnership.

12. *Communicate constantly.* In anything involving human beings, ongoing communication is crucial to success. Put every issue on the table and never make assumptions. That's a surefire formula for success.

The Chinese have a character in their written language that stands for the word *crisis.* There are two parts to this character. The first stands for the word *problem* and the second stands for the word *opportunity.* If the buyer does a good job of communicating needs to the supplier, and the supplier does a good job of communicating needs to the buyer, they can mutually work out all of the opportunities before they become problems. In doing so, they can look forward to a long, happy, and mutually beneficial partnership relationship.

SUPPLY PIPELINE MANAGEMENT

Up to this point, we have discussed the idea of partnership arrangements with suppliers. Companies are also starting to develop similar relationships with their customers. This allows them to tie customers and suppliers together in a "pipeline" that has significant advantages for all parties involved. To show how such a pipeline works, let's look at an example involving an engine manufacturer.

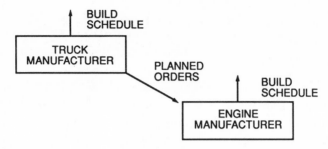

Figure 9.1

Before striking up the partnership shown in Figure 9.1, the engine manufacturer had to forecast how many trucks the truck manufacturer would build, and which engine models would be needed. This forecast was always incorrect in terms of quantity, timing, or mix. The errors caused the engine manufacturer to constantly change its schedules, which in turn became a headache for all the suppliers, so the engine manufacturer approached the truck manufacturer and suggested that their overall service and delivery could improve if everyone could work more closely together as a team.

Specifically, the engine maker suggested that the truck manufacturer install an MRP II system in order to get better control over their schedules and then share the planned orders for engines with the engine manufacturer. The truck manufacturer agreed, installed an MRP II system, and then began sharing a supplier schedule showing the next twenty-six weeks' schedule by engine model. The engine manufacturer responded by putting together a build schedule (that is, master-scheduled those planned orders in their own MRP II system) to support the truck manufacturer.

Next, the engine manufacturer approached all its suppliers and suggested that they also install MRP II. Several suppliers either already had MRP II or agreed to install MRP II if they didn't have it. The engine manufacturer then began sharing a supplier schedule showing the next twenty-six weeks by item number needed to support the truck engine schedule. This supplier schedule was based upon the truck manufacturer's plan broken down into the items each supplier produced. The suppliers then put together a build schedule to support the engine schedule and ultimately the truck schedule. All three of these schedules (the truck, the engine, and the purchased items) were then tied together electronically. Now, whenever the truck manufacturer changes its schedule, all the suppliers see the changes on-line. Forecasting has been eliminated and everyone is working to support the truck manufacturer's schedule. Inventory has been reduced, productivity improved, and costs lowered at every level in the relationship. The suppliers and the engine manufacturer, in effect, synchronized their schedules to that of the truck manufacturer. (See Figure 9.2.)

At some companies, this supply pipeline concept is being used to schedule several layers down into the supplier network. One large equipment manufacturer, for example, had been purchasing a long lead-time assembly from a supplier. The lead time was long because the equipment manufacturer's supplier had to purchase a machined forging used in the assembly which, in turn, had a lengthy lead time. The buyer convinced all the parties to sit down together and work out an arrangement whereby the buyer would schedule all the levels in the supply pipeline, from the raw material to the assembly. The buyer now schedules, with a supplier schedule, four layers down the pipeline. The layers include:

Supplier	*Material*
Steel Mill	Steel Ingot
Forging House	Hammer Forging
Machine Shop	Machined Forging
Primary Supplier	Assembly

This has allowed the buyer to reduce the lead time by 75 percent, reduce inventory in the pipeline, and reduce costs by 5 percent on the purchased assembly.

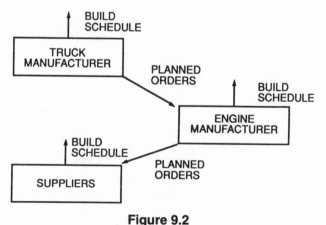

Figure 9.2

ELECTRONIC PURCHASING

As mentioned earlier, electronic purchasing allows the buying company and the supplier to link up directly. In a multilayered supplier pipeline like the one just covered, such communication is critical to quickly communicating the schedule to all levels in the pipeline. It is also critical for eliminating all the wasted time and costs associated with preparing purchase orders at each level in the pipeline. It makes good sense to use electronic purchasing to schedule all your suppliers. Two methods are commonly used to make the connection: electronic mail and electronic data interchange.

Electronic Mail

Electronic mail involves sending data or executable files from one computer to another via an information utility (such as MCI, the Source, etc.—more on these later). The data or files are actually sent to the recipient's "mail box." When the recipient checks his mail, he can retrieve the data or files and view them on-screen or print them out.

When used for supplier scheduling, electronic mail is a one-way transmission process in which you send an electronic message or file from one or more of your locations to the supplier's location(s). You can either compose a message through the information utility and then send it to the supplier, or you can use the information utility to download a spread sheet or word-processed document that you've created on your computer and transmit it to the supplier. The supplier can receive the message on a computer or a fax machine.

When we say "fax" machine here, we do not mean a system in which you type a purchase order and send a copy of it via electronic mail to your supplier's fax machine. That is the least desirable means of communicating information, because it still requires the preparation of a purchase order, which does little to reduce the paperwork. A more efficient way is to transmit the supplier schedule that appears on the PC screen directly to the supplier's computer or fax machine without the need to manually type a purchase order. Today, there are numerous fax "boards" that plug directly into a personal computer. These boards in effect allow you to use your supplier's fax as your printer.

Electronic mail offers three advantages over conventional correspondence: It costs less, it eliminates annoying and time-wasting "telephone tag," and it gives the buyer more time to buy. Let's consider each advantage individually.

1. *Reduced costs of purchasing materials.* The following cost comparison is based on 1989 average costs and will vary from company to company. Nevertheless, it will give you a good idea of the potential cost savings involved:

	Average Cost Per
Purchase order	$20–$30
Business letter	$10
Long-distance call	$2
Electronic mail:	
Up to 400 characters	$0.40
Up to 7,500 characters	$0.80

A character is defined as a letter, number, symbol, punctuation mark, and space. For example, "400 EA. PART #125 *DELIVERY 10/15/91*" contains thirty-five characters. Remember, even though no characters are visible between the letters and numbers, the spaces count as characters and are factored into the word count.

As you can see, electronic mail is considerably less expensive than conventional means of communicating purchasing information.

2. *No more "telephone tag."* One of the great advantages of electronic mail is that the two parties don't have to be at their computers at the same time. You can send your message at 3:00 A.M. while suppliers are sound asleep, and they can pick it up first thing the next morning. The software needed to tap into an electronic mail service varies in price from thirty to two hundred dollars, and the annual cost of a mailbox rental through an information utility is usually less than fifty dollars. These costs will be quickly paid back in terms of reduced long-distance calls and more effective communication with the supplier.

3. *More time to buy.* Ordering materials is much faster through electronic mail. When sending a purchase order to your suppliers through the U.S. Mail, you typically have to allow three or four days (or more during holidays) for the paper to reach domestic destinations, and up to two weeks to reach international locations.

In contrast, "electronic mail" plunks the purchase order in the supplier's electronic "in basket" in three minutes. This, of course, reduces total lead time for purchasing items and should allow the buying company to reduce its inventory by at least the amount of time it usually takes for the supplier to receive your purchase order through conventional means.

Finally, if set up properly, electronic mail not only eliminates the need for the piece of paper the supplier schedule is printed on, but it obviates the need for receiving reports and invoices. This in turn frees up some of the buyer's time presently spent doing paperwork, and allows him to do what he's paid to do: spend money well. See below for an example of electronic purchasing in action.

Electronic Mail Services

There are numerous suppliers of electronic mail services, some of the more familiar ones being MCI, Western Union, AT&T, Dialcom, Compuserve, and Digital Equipment. You simply call them and they will sell you the software you need and assist you in setting up the electronic mailbox network. While this will require you to do all the work in terms of installing the software and making the arrangements to get your suppliers hooked up to the network, there are companies called third-party suppliers that can do all the work for you. These companies have the necessary communications software on their computers and are already hooked up to an electronic mail service, such as Dialcom. For a fee, they will set up all the standard inexpensive RS232 (ANSI 12) computer-to-computer interfaces for you and bill you a per use charge. While the cost of using third-party suppliers is obviously higher than if you did it yourself, it may be well worth the added expense for several reasons. First, these companies have the experience to do it right the first time. Second, they eliminate your having to get involved in the software issues. Finally, they can get you up and running much faster.

Example of the Benefits of Purchasing Via Electronic Mail

The following actual example will further put the benefits of electronic purchasing in perspective, in terms of inventory investment and cost reduction. A medium-sized Midwest company was purchasing $8 million in steel annually direct from mills, and had $2 million in inventory on hand (four inventory turns). This company averaged a hundred and forty hours per week of downtime, because it never had the correct steel items on hand. The company recognized that it had two problems: too much inventory and too many purchased parts shortages. The company went to a local steel service center and negotiated daily deliveries of all the steel part numbers through an electronic purchasing supplier schedule that was based on its MRP system. (See Figure 9.3.)

Note that the first five columns of requirements are displayed in daily increments across the total schedule horizon of seven weeks. Every day, the schedule would be updated so it would show the next five days' requirements. Thus, the first schedule would show Monday through Friday of Week 1. The schedule generated the next day would show Tuesday of Week 1 through Monday of Week 2. The following day's schedule would show Wednesday of Week 1 through Tuesday of Week 2.

Each day, the supplier was shown the next five days' actual requirements plus planning data for the next seven weeks. These supplier schedules were sent electronically to eliminate any delays that could be caused by the regular postal service. The company was able to reduce its steel inventory to two days on hand, from $2 million to $50,000—a reduction of 98 percent. And because the company brought in steel daily to meet its needs, it was able to drop the downtime for lack of parts to fifteen hours a week—nearly a 90 percent reduction.

As a result, the company was able to improve service to the shop floor, reduce floor-space needs for warehousing the extra steel, free up $1,950,000 in cash from the inventory, which could be invested in the business, and save $390,000 annually in the cost of carrying the inventory (based on carrying costs of 20 percent).

Electronic Data Interchange (EDI)

EDI refers to computer-to-computer exchanges of intercompany business documents and information. In contrast to electronic mail, this form of electronic purchasing is a two-way transmission of data, and is often based on interactive systems that "talk" to one another. Three versions of EDI are in use today: One uses the supplier's computer, the second uses the buying company's computer, and the third uses both computers.

Supplier's Computer Provides Information
This EDI technique involves using on-line, real-time systems for remote order entry and inventory status reporting. Typically, the supplier provides the buyer with a terminal or software specifically written for the interface, which the buyer can use with a personal computer.

All tagged orders (*) are firm
Other orders are expected dates and quantities

Part Number	Description ECN No.	ECN Date	Buyer	Rec'd Last Week	Past Due	10/21	10/22	10/23	10/24	10/25	Week of 11/01	Week of 11/08	Next 4 Weeks
167458345	Plate Steel 1337A	03/14/81	030	6245		1250*	1250*	1250*	1250*	1250*	6300	6400	25600
167458407	Plate Steel 1337A	03/14/81	030	2091	34*	425*	425*	425*	425*	425*	2125	2075	8200
167458456	Plate Steel 1337A	03/14/81	030	880		440*	440*	440*	440*	440*	800	1320	4400
167458554	Plate Steel 1548B	06/23/84	030	3488	122*		1805*	1805*	1805*		3610	3600	14400
168122764	Plate Steel 1232A	11/21/80	030	84						84*	85	85	360

Figure 9.3

With this arrangement, the buyer can "look" into the supplier's inventory system and, if the parts are available, simply place an order for the parts on-line. The supplier will then ship the parts immediately from inventory. Some of these systems allow the buyer to actually view the supplier's future production schedules and allocate production not yet sold to match the buyer's requirements. The order is confirmed on-line and displayed on the screen. If desired, a printer at the buyer's site will then produce a confirming acknowledgment in duplicate, simultaneously printing the same order at the supplier's location.

The disadvantage of this type of a system is that if enough of his suppliers chose to have this kind of direct linkage, the buyer could end up needing an entire room filled with terminals in order to handle all the items from all the suppliers.

Buyer's Computer Provides the Information

This arrangement is the reverse of the preceding one; the buyer supplies the supplier with a terminal, so the supplier can look into the buyer's scheduling system on a daily or weekly basis. In some cases, the supplier actually does the purchase-part planning for the buyer and is required to advise the buyer in advance if he cannot meet the buyer's required delivery dates. Here's how it works. The supplier looks at the individual MRP planning reports for each part number he supplies. The supplier converts planned orders to scheduled receipts on the MRP planning report, which is a confirmation of the supplier's intention to deliver the parts on that day. If the date for the delivery is sooner or later than the need date in the MRP system, the MRP system alerts the buyer of the situation. The buyer then calls the supplier to determine why the delivery date is off.

When the supplier ships the parts, he enters the shipping information, along with the quantity and part numbers, into the computer terminal. This shipping information establishes an inbound schedule for the buyer's receiving dock. When the shipment arrives at the dock, Receiving enters the receipt on the system and the computer compares the receipt information with what the supplier said he shipped. The negotiated purchase prices are already in the database, and if the receipt matches the shipment information, the buyer's company pays for the parts upon receipt without the need for an invoice. This eliminates the need for a requisition, purchase order, receiving report, or invoice. As a result, the buyer has more time to buy and the company has eliminated a mountain of paperwork that adds no value to its product.

Buyer's Computer and Supplier's Computer Interact

The third version of EDI currently in use, and one that represents excellent opportunities for further development in purchasing, entails an interaction between the buyer's computer and the supplier's computer. Nissan's automobile plant is a classic example of this type of scheduling arrangement. The auto maker and 10 percent of its suppliers have MRP systems, which are all hard-wired together.

When an automobile body enters the paint booth at the manufacturer's facility, MRP identifies the customer, the options, the color, etc., and then schedules all downstream operations to assemble the proper options for that customer. At the same time, it downloads the schedule into the supplier's computers that are hard-wired into the system. Thus, the automobile maker not only schedules its own manufacturing operations, but, in effect, schedules its supplier's operations as well.

The suppliers manufacture their respective purchase parts and deliver them directly to the assembly line at the automobile plant. The parts are installed in the automobile three hours after they were scheduled in the supplier's computer by the buyer's computer.

Each of the suppliers hard-wired into the system delivers multiple shipments of purchase parts per day. As a result, this plant averages less than one day of on-hand purchase parts.

Figure 9.4 below demonstrates how this scheduling works. When the car enters the paint booth, its options are defined for all the suppliers.

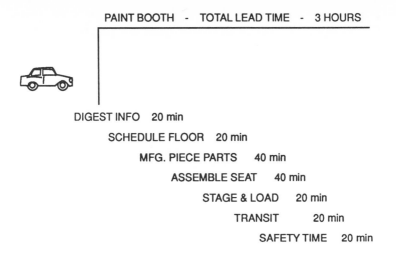

PAINT BOOTH - TOTAL LEAD TIME - 3 HOURS

DIGEST INFO   20 min
SCHEDULE FLOOR   20 min
MFG. PIECE PARTS   40 min
ASSEMBLE SEAT   40 min
STAGE & LOAD   20 min
TRANSIT   20 min
SAFETY TIME   20 min

Figure 9.4

Let's take the seat supplier as an example. The seat supplier knows that he is dealing with a brown vinyl four-door car with a specific model number. It takes the supplier's computer twenty minutes to digest that information and another twenty minutes to actually break it down to on-line schedules on the manufacturing floor. It takes the seat manufacturer forty minutes to make the piece parts for that particular seat and another forty minutes to actually assemble the seat.

In two hours, the seats are fully assembled, ready to ship. The car manufacturer wants to unload the seats directly out of the truck into the car on the assembly line, so the seats have to be staged and loaded into the truck in reverse order. In doing so, the first seat off the truck goes into the first car coming off the line. It takes twenty minutes to stage and load the truck and another twenty minutes to transport the seats from the supplier's assembly plant to the car assembly line. In this case, the seats arrive at the line about twenty minutes before they are put into the car. No requisitions, no purchase orders, and no receiving reports are necessary. Also, because the car manufacturer scheduled the seats and knows how many of each style they scheduled, no invoice is necessary either—it pays upon receipt.

This kind of timing would be impossible without EDI and computer-to-computer communication.

SUMMARY

- Prior to developing partnership arrangements, buyers and suppliers often have adversarial relationships.

- Partnership arrangements are characterized by a sense of teamwork, cooperation, and a shared goal of continuous improvement.

- The guidelines for effective partnerships should serve as the backdrop for all of your supplier scheduling efforts.

- Many companies are moving to supply pipeline partnerships that allow them to tie customers and suppliers together.

- Electronic mail is a one-way transmission of data from your location to the supplier's location.

- Electronic Data Interchange is a two-way transmission of data and refers to computer-to-computer exchanges of intercompany business documents and information.

Chapter Ten

Profile of a Good Buyer

Developing excellent suppliers to support an MRP II and JIT/TQC process means more than scheduling a "Supplier Day," or mailing out special memos that direct suppliers to participate in new product design, or instructing them to deliver future shipments in small lots to the assembly line just when needed. Unfortunately, that is all that happens in many companies. Today, there is very little incentive for suppliers to change their relationships with customers. As buyers come up with new programs, the suppliers see it as a one-way relationship that amounts to "I have to change, but the buyer does not," "I have to give up something, but I get nothing in return." And if the supplier does not go along with the new program, there is a veiled threat of potential loss of business. The relationship becomes adversarial, and the supplier becomes the villain because he will not support the program. The search then begins for a new supplier who will give the buying company what it wants.

To succeed in developing excellent suppliers, buyer and supplier must engage in a win-win relationship. That means the buyer's company must change the way it views its suppliers, must make changes in the way it schedules its suppliers, and then show and train its suppliers in MRP II and JIT/TQC. When companies work to combine the strengths of their suppliers with their own internal strengths, a mutually beneficial and long-term partnership merges. Chapter 2 described the attributes of a good supplier; this chapter covers the buyer's responsibility in creating a win-win arrangement.

QUALITY AND RELIABILITY

There are four quality and reliability requirements on the buyer's side. The first is a well-defined specification. This specification should define the minimum requirement of the item as simply as possible. If you do not need a tolerance or a certain requirement, it should not be on the specification. This only adds to the cost of the item you are buying and makes it more difficult for the supplier to conform to your spec. The specification should also take into account the discipline of design for manufacturability. (See Chapter 4.) In other words, the part should be designed in such a way as to eliminate areas of potential quality and reliability failures.

The second requirement is having a measurement system in place that identifies the defects in the incoming material. Ideally, you want to eliminate all incoming defects and certify the quality of all your suppliers, but achieving this level of quality requires a lot of work. When material is received, Quality Control should pull and inspect a sample of the shipment. All defects and the reasons for defects should be noted. This report then serves as the basis for determining the cause of the defect. Was it the supplier's process, equipment, employees, or suppliers that caused the defect to occur? What needs to be done to eliminate the cause so the problem does not occur again? The buyer must have a measurement system in place to identify defects if we are to eliminate them at the supplier's. A detailed discussion of quality measurement can be found in Chapter 12.

Third, with a well-defined specification and defect-measurement system in place, the buyer must be willing to become part of a joint approach to defect elimination. That means a willingness to send whatever resources are available to assist the supplier in the identification and elimination of the causes of the defects. If the defects stem from a tooling problem, can your tooling engineers assist the supplier's engineers in correcting the situation? If it is a problem of identifying how to eliminate the defect, maybe your quality-control people can assist. If there is a financial implication in solving the problem, maybe you can jointly share the cost. A good buyer will be agreeable to participating in the solution and not lay all the burden on the supplier.

Fourth, the buyer should be willing to train the supplier, if necessary, in various quality approaches. Today, the reality is that most suppliers do not have a meaningful Total Quality Control program in place and don't know how to go about setting one up. If the buyer's company has that expertise, it should be willing to share it with suppliers. If you do not have Design for Manufacturability, Statistical Process Control, or Total Quality Control in place, it is unreasonable to expect your suppliers to be using them either.

Finally, the good buyer will respect the supplier's expertise. Many of your suppliers are leaders in their industries, and the sooner you can tap that expertise, the better off the buyer will be. A good buyer will allow the supplier to participate in the original design of the part, which results in better-quality designs in a shorter time frame. Suppliers can anticipate tooling and equipment requirements, allowing them to get an early start on the long-lead-time items. They can anticipate quality and yield problems before the design is finalized. They can also make recommendations and participate in the identification of alternatives and trade-off analysis. This joint approach to defect elimination and supplier participation in the design process can provide excellent opportunities for building sound partnership relations with suppliers.

DELIVERY

A good buyer will give suppliers valid schedules. By "valid," we mean they are correct and attainable. In the traditional method of buying, the supplier is given a purchase order with a replenishment date. Both the supplier and the buyer know the date is earlier than when the first piece of the shipment is actually needed. Therefore, everyone knows the supplier can be late and the buyer will not run out of the material for a period of a few weeks. A good buyer will not only give the supplier valid need dates, but will continuously update the schedules as they change, via a supplier schedule, so the dates remain valid. The schedule must also be attainable in respect to the supplier's capacity. The buyer should not ask the supplier for a quantity of material in excess of the supplier's capability.

Beyond the short-term needs, a good buyer provides long-term visibility of needs to the supplier. If the supplier's quoted lead time is four weeks, the buyer should be willing to give the supplier visibility of needs for at least three months in the future. This would allow the supplier to plan raw materials, capacity, and equipment as productively as possible over time. The better the supplier is able to plan, the better chance the buyer's requirements will be met. (See Chapter 3 for a discussion of supplier scheduling and how need dates and future visibility can be given to the supplier.)

Time Fences

In addition, a good buyer has respect for the time fences established with the supplier. (See Figure 10.1.)

The typical supplier has three time fences that a buyer needs to respect. These are the time fences at which raw material is purchased, capacity is allocated on the manufacturing equipment, and the actual manufacturing lead time. Consider the manufacture of fabric, which was mentioned earlier. Although the supplier quotes ten-week lead times, the ten weeks is made up of four weeks to purchase the yarn, four weeks to weave the fabric, and two weeks to dye the fabric. The critical time fences are ten weeks and two weeks. If the buyer wishes to increase the rate of fabric deliveries on a weekly basis, it will take ten weeks for that to occur. If he wishes to change the color mix of the fabric on order, he can switch the colors at the two-week window.

A good buyer understands the supplier's time fences, schedules around those time fences, and does not expect the supplier to do something that is impossible. The good buyer would not ask for a color change one week out, because he knows that the dyeing process takes two weeks. A good buyer would not expect a rate increase in three weeks, because he knows that it takes the supplier six weeks to allocate capacity and weave and dye the fabric. Again, a good buyer respects the supplier's time fences and does not ask the supplier to do something that is impossible.

Delivery Specifications

A good buyer has well-defined packaging specifications that the suppliers have participated in developing and have agreed to meet. The buyer also negotiates all the terms and conditions in respect to the traffic

TIME FENCES

ACTUAL MANUFACTURING LEAD TIME	CAPACITY ALLOCATED ON MANUFACTURING EQUIPMENT	RAW MATERIAL PURCHASE LEAD TIME

TODAY

Figure 10.1

issues so that pickup and deliveries are convenient and economical for both parties.

Finally, a good buyer must be flexible. There are going to be problems in any relationship—things are not going to happen according to plan. The supplier's tool could break, he could have a quality problem, etc. The buyer needs to be flexible in either altering the schedule to work around the problem or assist the supplier in any way possible to get the problem resolved. In the old Greek wars, when the battle was lost, a messenger was sent back with the bad news. What happened to the messenger when the bad news was delivered? They killed the messenger. The buyer needs to avoid searching for the guilty party to blame when something goes wrong and look with the supplier for a joint solution. In a partnership arrangement, the supplier will notify the buying company as soon as possible if a problem is encountered, and the buying company will quickly move to help develop a solution.

PRICE

A good buyer enters into a partnership arrangement knowing that the buyer's company must share any risks jointly with the supplier should problems occur. If the buyer asks the supplier to build up inventory prior to a new-product introduction and the sales do not develop as forecast, the buyer may need to share in the cost of the buildup until it is worked off.

On the other hand, if a joint value analysis project reduces the item's cost dramatically, both sides should benefit. If the current cost of the item is $1.00 and the supplier makes a 10 percent profit, the buyer would pay a price of $1.10. Jointly, through a value analysis project, the cost of the material content of the item is reduced twenty cents. The current cost is now $.80 plus the 10 percent supplier's profit, putting the price at $.88 each. The buyer has netted a $.22 savings and the supplier has watched the profit drop from ten to eight cents each. How excited will the supplier get about the next value analysis project? Not very. If the relationship is to be a win-win situation, then the buyer must be willing to share in the risks and the savings with the supplier.

In addition, a good buyer will allow for fair pricing that allows the supplier to make a profit. Many buyers look only at the lowest price and constantly push to lower the price even more. If the supplier accepts the lower price in order to stay competitive, but is unable to lower the costs, he may be forced to cut corners to hold the ship afloat. This could cause delivery or quality problems in the future and would probably reduce the supplier's ability to be full-service, as described in Chapter 2. If you want a good supplier, you have to allow fair pricing.

Lastly, a good buyer will promptly pay the supplier for all shipments. Too often companies push their suppliers for quicker deliveries, better quality, and lower prices, and then pay them in forty-five to sixty days as a "reward." If you expect good service, if you want a win-win relationship, then prompt payment is mandatory.

RESPONSIVENESS

Just as the buyer wants the supplier to be very responsive to questions asked and to quickly follow through on problems and complaints, suppliers expect the same in return. The buyer should provide a list of the various personnel in the company (i.e., quality-control manager, technical services manager, tooling manager, supplier scheduling manager) along with telephone numbers where these people can be reached in the event of a problem. If the buyer is out of the office and the supplier does not know whom to call, the supplier may make a decision that is the opposite of what your quality-control manager might make. The supplier needs to be able to talk to someone who will understand the problem and provide a good answer.

In short, when the supplier calls, he should not get the runaround. This is part of working as a team dedicated to the quick resolution of problems.

LOCATION

If the buyer has multiple manufacturing locations, he should be honest about their capabilities and shortcomings. This will assist the supplier in understanding the various requirements at each plant and how to deal with the various messages received from each location. If one plant is significantly advanced in terms of SPC, JIT, and scheduling, it will be wanting more frequent shipments in smaller lot sizes, and will be interested in discussing the possibility of working toward certified quality. Another plant may be looking for truckload pricing only and may not be interested in a closer relationship with suppliers. The good buyer will be candid and open in communicating these differences and will work to assist the suppliers in developing the correct working relationship with each location.

TECHNICAL CAPABILITIES

Often, the buying company will have developed certain technologies that have allowed it to become a leader in its field. They may have developed certain tests that are used to measure quality and performance of their products that no one else has developed. The buyer's company should be willing to transfer certain of these technologies to suppliers to allow them to produce better products and to better understand the requirements of the items. One producer of high-quality, cutting-edge electronic devices, Tektronix, had developed several pieces of very specialized test equipment for monitoring its products. These pieces of equipment were developed and manufactured in-house and were not commercially available. Their suppliers did not have the same level of sophisticated test equipment, so they were not able to detect many of the defects that the buyer's company claimed existed. Realizing the situation, the buyer duplicated its test equipment and provided it free of charge to suppliers. As a result, the suppliers were able to track down the causes of defects. Today, 95 percent of the suppliers are certified and deliver direct to the production lines.

Research and Development

Many companies spend a significant amount of money on research and development of new products. Six months to a year after the development process starts, specifications for the purchase items are finally developed and sent to the suppliers for quotation. If the supplier is unable to produce the item as designed or has some suggestions for improving the item, Engineering then has to decide whether to delay the introduction or go with a less than desirable item.

The good buyer will get the supplier involved very early in the design process. Experience has proved that early involvement of suppliers, who are experts in their commodity lines, will result in better-quality designs in a shorter time frame. Suppliers can anticipate tooling and equipment requirements, allowing them to get an early start on long-lead-time items. They can also anticipate quality and yield problems before the design is finalized, and as team members they make recommendations and participate in the identification of alternatives and trade-off analysis. A good buyer not only asks the supplier to participate in the early phases of concept/design of the new product, but also provides that supplier with pertinent information such as pricing targets and benchmark information on quality reliability. This early supplier involvement provides excellent opportunities to cement sound partnership relationships with suppliers.

TWO-WAY STREET

As we said at the beginning of Chapter 2, a partnership is a two-way street. The relationship must be open and honest in terms of information. It must be based on a commitment to available resources, and it must share the risks as well as the rewards. Therefore, the buyer and the supplier must change the way they do business today if the relationship is to be win-win and long-term.

SUMMARY

- Developing excellent suppliers means more than scheduling a Supplier Day and mailing out special memos.

- To develop excellent suppliers, the buyer's company must change the way it views its suppliers, must make changes in the way it schedules its suppliers, and then show and train the suppliers in MRP II and JIT/TQC.

- A good buyer will be agreeable to participating in the solution and not lay the burden on the supplier.

- A good buyer will give suppliers valid schedules, not ask the supplier to do something that is impossible, and respect the supplier's expertise.

- A good buyer is flexible and realizes that problems are going to occur.

- A good buyer realizes that a partnership is a two-way street and is willing to share the risks as well as the rewards.

Chapter Eleven

Supplier Scheduling in Special Situations

Up to now, we have discussed supplier scheduling only as it applies to somewhat routine production items. This chapter will cover special applications of this technique. First we will address the topic of buying "scarce" commodities (such as integrated circuits). Then we will cover some of the issues involved in small companies and in "process" and "repetitive" manufacturers. This chapter will also discuss using supplier scheduling in a distribution company as opposed to a manufacturing firm. We will conclude with an overview of planning and scheduling Maintenance, Repair and Operating (MRO) supplies.

Buying Scarce Commodities

Depending on business conditions, all commodities can become difficult to acquire from time to time. Lead times for integrated circuits, for example, can vary dramatically over a year's time, creating special problems. As the demand for the commodity increases, its lead time may lengthen dramatically. Individual items within the commodity, as well as the commodity as a whole, may go on allocation. (See Figure 11.1.)

The supplier has a certain capacity, represented by the bottom of the funnel. The maximum rate at which the supplier can produce is shown by the rate of fluid flowing through the bottom. Obviously, if orders come in at a rate greater than the supplier's capacity, the level of orders in the backlog will increase and the supplier's lead time gets longer. At a certain level, represented by the top of the funnel, it becomes nearly

LEAD TIME FUNNEL

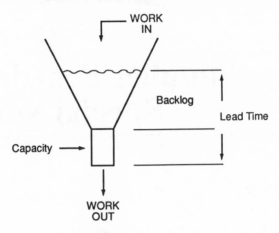

Figure 11.1

impossible for the supplier to put more orders into the backlog. At that point, the supplier will probably put all customers on allocation. Everyone is forced to take a percentage of what they might otherwise need. The funnel is now as shown in Figure 11.2.

The allocation can be very serious for a company if it is less than its real needs. An excellent way to attempt to avoid this situation is to work out an agreement with the supplier to set aside a certain percentage of

LEAD TIME FUNNEL

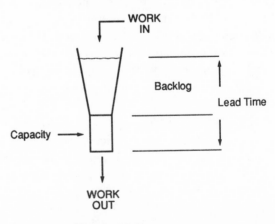

Figure 11.2

their capacity before the allocation situation occurs. Let's suppose that amounts to four hours a week of the capacity of a particular piece of equipment. If the supplier runs it for twenty-four hours a day, seven days a week, the total output from that piece of equipment is a hundred and sixty-eight hours. Using the capacity planning information, the buyer needs to contract for 2.4 percent of that capacity each week. That is represented by the thick line at the bottom of the funnel. (See Figure 11.3.)

Now it makes no difference if the supplier's quoted lead time goes from six weeks to sixty. The buyer is still able to schedule 2.4 percent of the supplier's capacity on a weekly basis. Even if the supplier goes on allocation, the company should get 2.4 percent of that equipment's capacity because of the contract the buyer has negotiated. That "scarce" commodity is now much easier to manage. Based upon individual needs for the items the buyer has scheduled capacity for, the buyer always counts on getting the four hours of capacity each day, each week, or each month from the supplier.

Be aware, though, that this approach to scheduling capacity and selecting out of that capacity what is needed is not a panacea. It will take a lot of work, in some cases, to understand how the supplier schedules the item and how many work centers are involved at the supplier's facility. If multiple work centers are involved at the facility, this approach may be difficult to negotiate with the supplier. But if a capacity

LEAD TIME FUNNEL

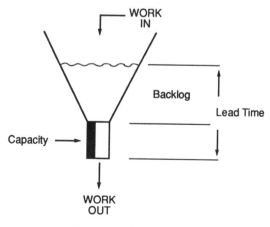

Figure 11.3

reservation can be negotiated, supplier scheduling will be much easier for those scarce commodities.

SMALL COMPANIES

It is often claimed that supplier scheduling is a good concept for large companies but not for small ones. Buyers for smaller businesses may believe they have no "clout" with their suppliers and that the system therefore will not work the same way for them. This belief implies that clout is all it takes to get supplier performance—that unless someone buys in high volumes, the supplier really isn't interested in the business. This is simply not true in most cases. If it were, the large companies would never run out of any parts and would always get first call on supplier capacity. Yet almost everyone has heard about large companies that have missed their monthly production goals for lack of purchased parts.

It is not always clout that counts, it is the quality of the schedule that does. We have seen small companies with good supplier schedules get excellent supplier performance in all types of market conditions. In fact, the findings reported in Appendix B show that 80 percent of these companies surveyed purchased less than $30 million annually, with two companies spending only $4 million annually. These companies spent little with each individual supplier, but were able to get good service because they had valid schedules. When suppliers understand what the supplier schedule means and how they can use it to maximize their planning to meet customer needs, company size becomes far less important. Even small companies have found that supplier scheduling improves supplier delivery performance dramatically.

"PROCESS" COMPANIES

Companies in process industries, such as chemicals or paper, typically run on a continuous basis, and shutting down equipment, such as a glass furnace, is very expensive. All raw materials required must be continuously available. Since they are always running, such plants tend to use very large quantities of basic raw materials quickly. It is quite common to bring in tank cars or trucks of chemicals weekly or even more frequently. Often, the storage facilities for these basic chemicals cannot hold more than a few days' supply, so inventory turnover tends to be very high. Supplier scheduling fits well into this type of environment.

Most process plants have a daily production schedule, which becomes, in effect, the master production schedule in MRP II. They also have formulations or recipes (bills of material) for their products. Because of the need to maintain continuous production, process companies regularly verify the on-hand inventory of all raw materials. These elements, of course, are the key ingredients for Material Requirements Planning: a master schedule, bills of material, and the inventory status. By breaking the supplier schedule down into a daily schedule on purchased items, the Purchasing Department can schedule deliveries to arrive on the day needed. Here also, supplier scheduling handles the daily scheduling tasks and allows the buyer to spend time on negotiations, alternate sourcings, etc.

One potential variable in certain process companies is the yield of the end product. Process variables or process problems may cause the yield to differ from day to day. In other words, it may take more raw materials than expected to produce the total batch weight desired. This is somewhat analogous to the scrap some manufacturers in nonprocess industries experience on raw materials. If the company can forecast scrap or yield variances, they can be expressed in the bill of material. If they cannot be anticipated, the replanning feature of MRP will take these yield variables into consideration. If production takes more of a particular raw material than anticipated, the on-hand inventory will reflect this. MRP would then recalculate the need date based upon the lower inventory and advise the supplier scheduler accordingly. Therefore, the supplier schedule can reflect the true need date on each item bought, taking into account all the variables of the process.

Replanning is one of the strong features of MRP II. It not only plans the first time, but replans based upon the current situation. Reorder points and other ordering systems don't have this replanning capability. The MRP-generated supplier schedule in a process company can be a natural to support a continuous process.

REPETITIVE MANUFACTURERS

High-volume repetitive manufacturing is another natural for supplier scheduling. By definition, the same parts are being used over and over again, day in, day out. High-volume repetitive situations lend themselves readily to Just-in-Time scheduling, as we discussed in earlier chapters. In many repetitive environments, where the items required are fairly simple to manufacture, the supplier schedule can actually be used as the supplier's own production schedule.

As an example, suppose a company uses wood in its product and the only variable is the length of the piece. The buyer's company makes a thousand end items per week and requires the wood pieces to be either six inches, six and a half inches, or seven inches in length, depending upon which version of the end item is scheduled for production. The supplier would set aside the capacity required to make a thousand cuts on a table saw. The supplier schedule would be given directly to the operator of the table saw, and it would tell the operator how many of each length to cut each day or each week. The other alternative, as mentioned earlier, is to use a Kanban card or similar mechanism to break down the schedule on a daily basis.

Since the purchased parts are repetitive, the supplier schedule tends to be much more stable in these cases. This makes the supplier's planning of capacity and raw materials much more efficient and should allow better pricing stability. If a company produces twenty thousand chairs per week and each chair takes one yard of fabric, the supplier knows he needs the yarn and capacity equal to twenty thousand yards every week. The supplier is able, therefore, to do a better job of planning internally and to secure stable pricing and supply with their suppliers. This further assures their customers of a constant product at a consistent price over time. Supplier scheduling is well suited to highly repetitive companies. Figure 11.4 is an example of a supplier schedule in a repetitive manufacturing company.

DISTRIBUTORS AND RESELLERS

Distributors and resellers are different from the companies discussed so far in this book. With few exceptions, they do not manufacture any of the items they sell. Typically, they purchase finished items from various manufacturers and resell them to retailers.

Distribution companies can develop supplier agreements for all the items they buy from a manufacturer, and then schedule in the items as they are needed. Sound familiar? That is what all buyers do, whether they are purchasing castings, fasteners, or finished items for resale, such as hammers or cosmetics or whatever.

Let us look at the sources of demand on a distribution company's buyer using Distribution Resource Planning, part of an MRP II system. In this example (see Figure 11.5), there are three regional distribution centers (DCs) and one central distribution center. The central DC

All tagged orders (*) are firm
Other orders are expected dates and quantities

Part Number	Description ECN No.	ECN Date	Buyer	Rec'd Last Week	Past Due	Requirements 10/21	10/22	10/23	10/24	10/25	Week of 11/01	Week of 11/08	Next 4 Weeks
167458345	Plate Steel 1337A	1X4X3/4 03/14/81	030	6245		1250*	1250*	1250*	1250*	1250*	6300	6400	25600
167458407	Plate Steel 1337A	1X5X3/4 03/14/81	030	2091	34*	425*	425*	425*	425*	425*	2125	2075	8200
167458456	Plate Steel 1337A	1X8X3/4 03/14/81	030	880		440*	440*	440*		440*	800	1320	4400
167458554	Plate Steel 1548B	2X4X1/2 06/23/84	030	3488	122*		1805*		1805*		3610	3600	14400
168122764	Plate Steel 1232A	2 ft Dia. 11/21/80	030	84						84*	85	85	360

Figure 11.4

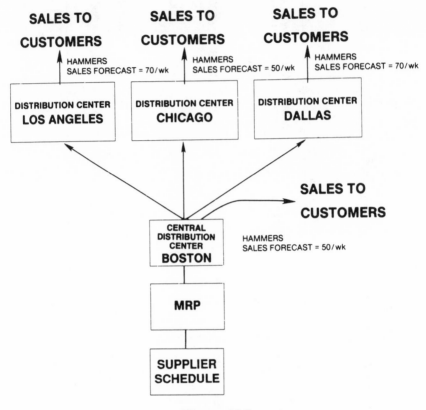

Figure 11.5

replenishes the regionals, which ship to customers. In addition, the central DC services customers in its area. Each regional DC would have a sales forecast and an on-hand inventory for every item and would also know what is in-transit from the central DC.

On the hammers it stocks, the Dallas DC has a forecast of seventy per week, a hundred and twenty currently on hand in inventory, and two hundred in transit from the distribution center due next week (Period 1). (See Figure 11.6.) Distribution Resource Planning would calculate when Dallas will need more hammers (planned orders).

Dallas will need a shipment of two hundred hammers in Week 4 and another shipment of two hundred hammers in Week 7, to avoid going below the safety stock of one hundred. Given the two-week replenishment lead time, the central DC will need to ship to Dallas in Weeks 2 and 5; hence the planned order releases of two hundred in each of these weeks. These Dallas planned orders are input to the central DC's Material Requirements Planning system for the hammer, along with the planned orders from the other two DCs and the regional distribution center.

Note that Los Angeles orders in quantities of two hundred and Chicago in quantities of one hundred. DRP then calculates a schedule of the total requirements for hammers to be delivered from the manufacturer to the distribution center. (See Figure 11.7.)

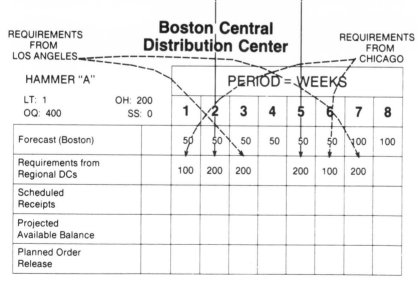

Dallas Distribution Center

HAMMER "A" LT: 2 OQ: 200 OH: 120 SS: 100		PERIOD = WEEKS							
		1	**2**	**3**	**4**	**5**	**6**	**7**	**8**
Forecast		70	70	70	70	70	70	70	70
Scheduled Receipts		200							
Projected Available Balance	120	250	180	110	240	170	100	230	160
Planned Order Release			200			200			

Boston Central Distribution Center

REQUIREMENTS FROM LOS ANGELES

REQUIREMENTS FROM CHICAGO

HAMMER "A" LT: 1 OQ: 400 OH: 200 SS: 0		PERIOD = WEEKS							
		1	**2**	**3**	**4**	**5**	**6**	**7**	**8**
Forecast (Boston)		50	50	50	50	50	50	100	100
Requirements from Regional DCs		100	200	200		200	100	200	
Scheduled Receipts									
Projected Available Balance									
Planned Order Release									

Figure 11.6

Boston Central
Distribution Center

HAMMER "A" LT: 1 OH: 200 OQ: 400 SS: 0	PERIOD = WEEKS							
	1	2	3	4	5	6	7	8
Forecast (Boston)	50	50	50	50	50	50	100	100
Requirements from Regional DCs	100	200	200		200	100	200	
Scheduled Receipts								
Projected Available Balance	50	200	350	300	50	300	0	300
Planned Order Release		400	400			400		400

(Note: Projected Available Balance OH: 200)

Figure 11.7

The planned order information for Hammer A at the central distribution center is consolidated via MRP onto the supplier schedule along with planned orders for all other items purchased from that supplier. The supplier scheduler at the central DC would review the schedule and forward it to the supplier.

If the hardware distributor in this example were purchasing hammers from more than one manufacturer, for example, both Stanley Tool and Tru-Temper, then each of their respective supplier schedules would list their portion of the total hammer requirements. The buyer has negotiated master purchase agreements for hand tools, which includes the hammers just planned. The supplier schedule can be used to schedule the hammer and all the other hand tool deliveries to the distribution center.

MRO ITEMS

The typical Maintenance, Repair and Operating (MRO) buyer is drowning in paperwork. Consider that if he buys twenty thousand items an average of three times a year, that's sixty thousand requisitions, sixty thousand purchase orders, and sixty thousand receiving reports. It is no wonder the MRO buyer has little time to do anything but shuffle paper. Because of this time constraint, the buyer typically finds it hard to do systems contracting or value analysis on the MRO items.

These items, however, can be handled and scheduled in much the same way as production items. To schedule any item on MRP, three things are needed: a master schedule on the end item, an inventory record on each item, and a bill of material that includes every item to be scheduled. Let us look at the bill of material.

First, every item must have an item number so the system can keep track of it. This may be a problem for some companies that do not currently have item numbers assigned to MRO items. Those companies would have to assign item numbers for all the MRO items that they want to schedule on MRP II.

The second requirement of the bill of material is the quantity of each item to be scheduled. For example, suppose that a piece of consumable tooling, such as a drill bit used in manufacturing an item, needs to be replaced. On average, the drill bit is replaced after producing ten parts, so the quantity per for that drill bit would be one-tenth. For every ten parts required, the MRP II system will require one drill bit.

The third requirement of the bill of material is a structure showing how the item is used. This structure is represented by Figure 11.8.

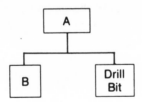

Figure 11.8

If the job demands a hundred of Part A, the system would plan a requirement for ten drill bits and one hundred undrilled parts (B), then schedule the drilling work center to complete the drilling operation. Likewise, if the drilling operation requires cooling oil, that could be scheduled by making it a component of Part A as well. See Figure 11.9.

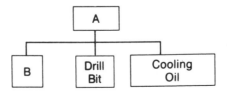

Figure 11.9

By including the cost of the consumable tooling or the operating supplies in the product costs, a company can get a more accurate total cost of producing a product. On the other hand, some companies may not want these expense items in the product cost. By setting the costs of the drill bit and cooling oil at zero, they can be scheduled using MRP II and not affect the product cost.

An inventory record on the drill bit and cooling oil is required to determine the starting on-hand balance. If there are eight drill bits on hand and the MRP II system requires ten, it would indicate that at least two more must be acquired to meet the production schedule. The master schedule, the schedule of the end items to be built on a weekly or daily basis, would then drive the week-by-week drill bit requirements. These drill requirements would then be scheduled from the supplier on the supplier schedule. The buyer would negotiate a contract on drill bits, and the supplier scheduler would release the individual weekly quantities to the supplier. In this way, the operating supplies would be scheduled exactly the same as the production items.

Here is an interesting twist on organization. In one company we know, the supplier scheduler for consumable tooling—drill bits, saw blades, reamers, etc.—is a member of the Industrial Engineering Department. In another company, the supplier scheduler for packaging material is in Shipping. Good basic systems (knowing what is needed and when) and short, clear lines of communications (supplier scheduling) permit organizational flexibility, one aspect of which is being able to put functions "where the action is."

The maintenance items for the equipment in the factory can also be supplier scheduled. Many companies have a regular preventive maintenance schedule. This can be included in the master schedule. Each piece of equipment is given a number, such as Press 1 (P1), Press 2 (P2), and so on. When the maintenance staff plans the preventive maintenance on an individual piece of equipment, they intend to replace certain items. These items are then structured into the bill of material as shown in Figure 11.10.

If the lead time on each of these items to be used is one week, and the maintenance on Press P2 is scheduled for the week of July 15, MRP would schedule the items needed from the supplier releasing the orders in the week of July 8.

This approach can also be adapted to work for items that aren't replaced consistently. Suppose that when the company does preventive

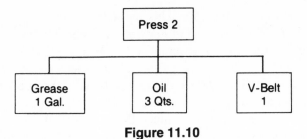

Figure 11.10

maintenance on a piece of equipment, it's not always necessary to replace the V-belt. By checking the maintenance records, it is possible to see that it is necessary to replace it only 80 percent of the time. The company can put a .8 fractional quantity in the bill of material. MRP will then compensate for this and schedule in the correct quantity of the V-belts.

By inputting the annual preventive maintenance schedule for all the equipment in the factory into the master production schedule, preventive maintenance parts can be placed on a supplier schedule. Now that the buyer knows the annual requirements for oil, grease, V-belts, and so on, the buyer can negotiate annual contracts on these items with releases as required. Like the production item buyer, the MRO buyer is spending time contracting, not handling pieces of paper.

This technique can be extended to office supplies, hand tools, pipe fittings, ad infinitum. Here is an illustration. Most companies have a central stockroom that contains all the office supplies. Every item in the stockroom can be assigned an item number, which could, if desired, match the office supply supplier's catalog number for every item. An inventory record is created and a minimum inventory established for each item (such as one dozen ball-point pens). When someone needs a pen, it is issued from stock. When the inventory drops below the minimum (in this case, to eleven), the MRP system schedules a replacement shipment from the supplier in an appropriate lot size. If the company inputs a forecast on each item based on either expected future usage or on historical usage, it is possible to get a time-phased requirements plan like that for all the other items being scheduled on MRP II.

This idea works equally well for in-plant construction. A simple example would be to build a wall to divide a conference room. A work order would be generated and assigned a number, such as WO137. The maintenance manager would develop a "parts list" of the materials needed to complete the wall. This parts list would be input to the computer in the form of a bill of material. The end item number would be the same as the work order number. See Figure 11.11.

Purchasing has negotiated a "list-less" contract with the local lumberyard covering all purchases for the year (that is, the published price on all the items, an across-the-board discount on all purchases). The Maintenance Department determines that they can do the job in six hours, and schedules the job for September 15. The lead time on the purchased items is two days. On September 13, the supplier schedule triggers the delivery of the various items required from the lumberyard. On September 15, the maintenance dispatch list would schedule the wall to be built. When the wall is completed, the work order is indicated as complete and the one-time bill of material is removed from the system. Again, the supplier schedule has been utilized to bring in the purchased parts on time.

Permanent tooling—both new tooling and preventive maintenance of existing tooling—can be scheduled in much the same way. By structuring a bill of activities (see Figure 11.12), it's possible to schedule all the activities necessary to have the new tool on the date required. When the tooling request is completed on June 1, the operation is reported as complete. The MRP system then schedules the Design Engineering Department to complete the design, a draftsman to complete the drawing specification, and lastly, the actual building of the tool.

MRP II is fundamentally a network scheduling system. Project activities of a one-time nature, such as designing and building a tool, typically scheduled by PERT (Project Evaluation and Review Technique), CPM (Critical Path Method), or other scheduling systems, can be scheduled via MRP II. A planning bill of material, or bill of activities as in Figure 11.12, is used instead of the PERT or CPM chart. Once estab-

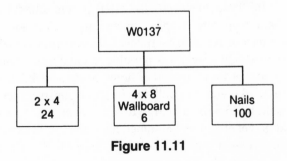

Figure 11.11

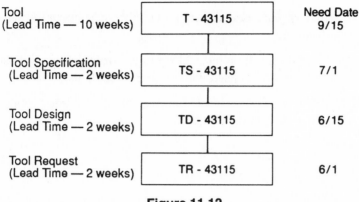

Figure 11.12

lished, this planning bill of material or bill of activities is entered into the computer just like a standard bill of material. Then, as each step is completed, the MRP system schedules the next operation in time to meet the completion date of the project. Companies have even erected buildings using MRP II scheduling techniques. MRP II has the capabilities to schedule all of the resources needed in a company and shouldn't be seen as just manufacturing resource planning.

MRP II and supplier scheduling works as well for the distributor and reseller as the manufacturer, in large companies and small, process, repetitive, or one-of-a-kind manufacturing companies. It represents a universal solution to a universal problem in purchasing: too little time to do a professional buying job. Supplier scheduling is limited only by the user's imagination and perception of what it can do.

SUMMARY

- An excellent way to manage scarce items is to work out an agreement with the supplier to set aside a certain percentage of their capacity.

- When it comes to excellent supplier performance, it is not always clout that counts, it is the quality of the schedule that counts.

- Supplier scheduling works well in a process type of environment.

- Replanning is one of the strong features of MRP II.

- High-volume repetitive situations lend themselves readily to Just-in-Time scheduling.

- Distributors and resellers can use supplier scheduling the same as manufacturers.

- MRP II can free the typical MRO buyer from the massive paperwork normally associated with buying these items.

- Supplier scheduling is limited only by the user's imagination and perception of what it can do.

Chapter Twelve

Performance Measurements

Many factors go into the successful use of MRP II and JIT/TQC. Two important ones are performance measurements and a commitment to continuous improvement. Performance measurements provide a window into areas that can be improved upon in the future. In successful companies, continuous improvement becomes a way of life for employees, suppliers, and customers. It is also a key tool in maintaining a competitive advantage in today's manufacturing and purchasing environments.

PERSPECTIVE ON MEASUREMENTS

If MRP II is working correctly, need dates are being given to the supplier by either the supplier schedule or a Kanban card. If education has been done properly, users should start noticing a marked improvement in supplier performance. The only way to be absolutely certain of real progress, however, is to measure that performance.

Going by the Numbers

The first step is to establish objective measurements. The buyer should sit down with the supplier, set realistic goals, and develop a timetable for reaching them. These goals cannot be stated in subjective terms like "do better" or "work harder." This just leads to great philosophical discussions about what "do better" means and whether the supplier did in fact "do better." It leaves open to subjective opinion whether, in fact, the performance did improve. But if the goal is set in objective, numerical terms, the answer to the question "Did performance improve?" is obvious. If the goal was 95 percent and the supplier improved from 90 to 93 percent, the performance did improve but is still short of the goal. As the saying goes, "If you don't know where you are going, any road will get you there." The objective of performance measurements is to establish which road you want to take, where you currently are on the road, and where you ultimately want to end up.

Companies often need to make some basic changes in the way they reward performance. For example, if two buyers receive the same percentage pay raise at the end of the year, when one had suppliers who were 70 percent on-time and the other 98 percent on-time, the incentive to perform to the maximum level is diminished. Compensation systems, therefore, should reward people for good performance. The same principle is applicable to suppliers. If the 70 percent on-time supplier receives the same percentage of the business as the 98 percent on-time supplier, there is little motivation to work harder and improve performance. Both buyers and suppliers must realize that they are expected to meet certain standards of performance, and should be made aware of the unpleasant consequences of missing their targets.

Goals should also reflect current realities. If a company has just begun measurement reporting and finds that a particular supplier's on-time delivery performance is 70 percent, it is neither fair nor realistic to expect that supplier to achieve 95 percent by next month. Measurements are a starting point—they help people learn how to improve performance by pointing out where they are deficient and by establishing achievable timetables to reach desired levels.

Measurement reporting represents an opportunity to develop a better understanding between the supplier and the buyer of each other's needs. In this way they can be used as a basis for improved performance. If measurements are viewed simply as a punishment/reward tool in supplier relationships, the results may be just the opposite of those desired.

Small Steps Versus Giant Leaps

Progress toward performance goals should be monitored. This is done with a process that entails setting up intermediate goals as yardsticks for improvement. Let us say Supplier A is currently 70 percent on-time, and you would like him to achieve 95 percent on-time performance six months from now. This supplier agrees that six months is a realistic time frame. Your intermediate goals are 80 percent in two months and 90 percent in four months. Supplier B, on the other hand, is 85 percent on-time. Since your goal is to have all suppliers at 95 percent on-time six months from now, and Supplier B also agrees that that length of time is realistic, his benchmark would be 90 percent two months from now and 93 percent four months from now. This approach allows you to measure all of your suppliers' improvements in performance on a regular basis, taking into account where they are today, with confidence that they will improve steadily and hit your 95 percent on-time goal within six months.

One other perspective on measurements is needed before we look at how results are measured. We normally talk about goals in terms of 95 or 98 percent attainment, i.e., quality is 98 percent (98 out of every 100 lots received are accepted, while only 2 are rejected). Based on most companies' current level of performance, those could be very challenging goals. A nationwide survey of purchasing personnel, reported in *Purchasing Magazine*, revealed an average on-time performance of 82 percent and in-specification of 92 percent. Therefore, 98 percent sounds very good to most people in purchasing today.

But consider the remaining 2 percent out of specification in the context of parts per million. A 2 percent defect level means you are getting 20,000 defects for every one million parts you receive. For certain, 20,000 defects is totally unacceptable on a continuing basis. This is why the continuous improvement process is so important. But to use it, you must first put a measurement system in place to determine where you are today. Next, you need to establish goals and programs to get the purchasing environment under control—95 to 98 percent in terms of performance attainment. Finally, you need to put into action a program to continuously improve your performance beyond those "acceptable" levels of performance.

If you are at 20,000 defects per million (2 percent), you could use a tool such as Pareto analysis to identify the reasons for the defects, solve the problem for some defects, and lower the total number to 19,000 defects per million. If you can get down to 19,000 defects, you can achieve 18,000, and so on. A leading manufacturer of copy machines was able to drop its level of incoming quality defects from 10,000 parts per million to 122 parts per million. Although this represents 99.9988 percent in-specification, the company still is not satisfied and vigorously pursues its continuous improvement program.

How Results Are Measured

To get control of your purchasing environment, you must first identify areas that need improvement. A good purchasing performance measurement system should cover eight different areas, including:

1. Delivery

2. Quality

3. Price

4. Lead times

5. Inventory investment

6. Schedule completions in the plant

7. Cost reduction/value analysis

8. Inbound freight cost reduction

If your company is already measuring other areas as well, combine them with the eight discussed below to provide an overall picture of supplier, buyer, and supplier scheduler performance.

Delivery
Because the company is displaying need dates rather than due dates on its schedules, the supplier must deliver on-time or a shortage will probably result. The supplier, the buyer, and the supplier scheduler must all realize that the dates on the supplier schedule are "drop dead" dates. Some companies tell their suppliers that a shipment will be considered "on-time" if it arrives anytime from seven days before the due date to seven days after it. That tells the supplier that the need dates are at least seven days later than the due dates. Performance cannot be measured honestly when an order shipped five days late is considered on-time. Initially in such a situation, it is preferable to allow suppliers to ship up to two weeks early. This still provides the same two-week delivery policy as the "seven days early to seven days late" policy but without violating the validity of the need date.

Once the need date is understood by suppliers, and suppliers demonstrate their ability to routinely deliver in that two-week window, the continuous improvement process can begin. The window is first moved from fourteen days to ten days. When the supplier is able to live with ten days consistently, the window is closed to seven days.

Leading MRP II/JIT companies today allow the supplier to ship up to two days early, no days late. And it is now becoming common practice in some industries to actually tell the supplier the *time of day* the shipment is wanted, not just the day it is wanted.

If the trucking time for shipping a product varies between three days and ten days, Purchasing can initially instruct the supplier to ship it ten days before the need date, so it will always arrive on-time (before the need date). Once this performance level has been accomplished, the trucking company can be included in the continuous improvement process. Ask questions such as "What can we do to cut down the variability in transit times?" and "Do we need to work with a specialty contract carrier to get the transit time to a consistent three days?"

The buyer should also define the past due cutoff date. If the order is due on Wednesday afternoon, the supplier must realize that a Thursday-morning delivery will be considered past due. If the order is for five thousand and the supplier ships only four thousand, the remainder of the order is also considered past due on Thursday morning. Suppliers must realize that companies are serious when they say an order must be delivered on-time. Once the ground rules for on-time delivery have been established, performance can be measured more objectively.

It may also be necessary to change the way on-time performance is calculated. Typically, in companies without supplier scheduling, on-time performance is measured by adding up all past due purchase orders and dividing that number by the total number of open purchase orders.

Unfortunately, this measurement technique has some major flaws. It is too general and does not reflect how well a particular buyer or supplier did in a particular week or month. Consider the following example. You have one thousand open orders in Purchasing with your suppliers and one hundred of them are past due. The current performance level would be 90 percent (one hundred past due divided by one thousand open orders). If the supplier's lead times were to increase, the buyer would need to release more orders further in the future to cover the longer lead times. Using this example, there may now be twelve hundred open orders. At the same time, the on-time measurement would increase to 92 percent (one hundred divided by twelve hundred). Is that because the supplier and the buyer are performing better? Of course not—it is purely a paper improvement. Obviously, this technique does not provide a true measure of performance.

Here is another example. A buyer has five open orders with a supplier, each for twenty pieces. (See Figure 12.1.) One of those orders is past due, so the supplier is 80 percent on-time (one order past due out of a total of five open orders).

Current Date 7/1					
Due Date	6/25	7/5	7/15	7/25	8/5
Quantity	20	20	20	20	20

Figure 12.1

On July 10, the supplier ships the twenty pieces due on June 25. (See Figure 12.2.) The July 5 shipment is now past due, however, and another order, due August 15, is added. The supplier is again 80 percent on-time, because one is late out of five that are due.

Current Date 7/10					
Due Date	7/5	7/15	7/25	8/5	8/15
Quantity	20	20	20	20	20

Figure 12.2

On July 20, the supplier ships the twenty pieces due on July 5. (See Figure 12.3.) The order due July 15 goes past due and another order due on August 25 is added. Look at the open order report of July 20—the supplier is once again 80 percent on-time.

Current Date 7/20					
Due Date	7/15	7/25	8/5	8/15	8/25
Quantity	20	20	20	20	20

Figure 12.3

The problem here is clear: The supplier has shipped every single order late and, in fact, is zero percent on-time in terms of delivery. A delivery performance system must measure true on-time performance. The system should track the receipt date against the due date on the supplier schedule and give the buyer and supplier an objective score on performance against the schedule for that week or month.

Figure 12.4 shows an example of a format for such a measurement report. The buyer and supplier would go over the report each month. Orders considered past due would be displayed below the totals column for the supplier's review. The principle of "Silence is approval" applies; unless the supplier disagrees with the monthly performance percentage, he is held accountable for it.

At this point, the continuous improvement process can begin. Part number 147 was received seven days late. By working with the supplier, the cause of the late shipment can be resolved. Your thinking and questioning should go something like this: "Was it a quality problem, tooling problem, scheduling problem, etc., that caused the part to be late? What can *we* jointly do to eliminate this problem from occurring again? If *we* can jointly eliminate the cause of the late shipment so it does not occur again, we should expect performance to improve toward our goal of on-time shipments." It is important to remember that you may have caused the problem by a late schedule change or engineering change that the supplier did not have enough time to react to. The key to continuous improvement is to identify the cause and prevent problems from occurring in the future. Finally, since each item is tied to a buyer, a supplier scheduler, and one or more suppliers, a similar measurement report could be generated for each of these people.

You do not need sophisticated on-line systems to get on-time measurements; a microcomputer and spread-sheet package will suffice. For that matter, you can collect and track the measurement data manually. When the shipment arrives, the personnel at the receiving dock can fill out a simple one-part form such as the one shown in Figure 12.5. This

---

ABC Supply Company
Month of July

Scheduled Receipts Due	125
Scheduled Receipts Past Due	1
On-Time Performance	99.2%
Order Past Due 200 Part #147	Due 7/15, received 7/22

Figure 12.4

DELIVERY

PART NUMBER _____

VENDOR NUMBER _____

DATE RECEIVED _____

LOCATION / DOOR _____

QUANTITY RECEIVED _____

UNIT OF MEASURE _____

Figure 12.5

information can then be either entered into the computer or manually checked against the open order file or scheduled receipts on the supplier schedules to determine whether the supplier was on-time or not. The shipment was either complete and on-time or it was not—the measurement is as simple as that.

If you only have one receiving door, obviously you would not need such information. But for companies with many receiving docks, it is a way to see if the supplier delivered it to the particular door you requested (Door 7, Building 14), and if not, why. In a continuous improvement effort where you are requesting the supplier to deliver product directly to the using department on the manufacturing floor, the measurement is a way of tracking his ability to do that. In that case, the department number would be input instead of the door number and the supervisor of the department would fill out the receiving form.

Reliable delivery is critical to MRP II. Since the supplier schedule is based on need dates, the goal should be a minimum 95 percent on-time. Now that suppliers are receiving the information they need to plan capacity and raw materials via the supplier schedule, they can reasonably be expected to meet the need dates. Those starting out at 70 percent on-time should be able to meet their immediate goals of 80 percent, then 90 percent, and then 95 percent. But don't stop there; 95 percent should be considered as the minimum acceptable level of performance. Why not 98 or 99 percent? Why not 100 percent?

The delivery measurement report will give Purchasing the information needed to identify the problem suppliers or items. When those problems are resolved, performance can be expected to improve. The challenge, then, is to make continuing improvement a way of life. The challenge is to get better and better. Some companies today are achieving nearly 100 percent delivery performance from their suppliers. They did not get there overnight; it has taken lots of hard work on the part of many people. But it is certainly worth it.

The survey in Appendix B indicates that Class A/B MRP II companies using supplier scheduling averaged 97 percent on-time supplier delivery performance. The average company took six months from the start of supplier scheduling to accomplish 97 percent on-time delivery performance. Those companies using a Kanban mechanism and receiving daily shipments from their suppliers averaged over 99 percent on-time delivery.

Quality

Out-of-specification material can represent just as many headaches as material that arrives late. Again, because supplier scheduling deals with need dates, the items must be of acceptable quality when they arrive at the receiving dock. It does not do a buyer any good to bring items in just in-time if the quality is bad. It is equally disadvantageous to bring in good-quality material late; both the delivery date and quality level are critical to supplier scheduling. Companies need to work with their suppliers to assist them in developing a total quality approach that includes statistical process control and defect-prevention tools that will assure quality items every time. That means you will need to work with your suppliers to be sure they have the necessary tools, test equipment, and fixtures to make a quality item. Well-defined specifications on every item and a written agreement to those specifications from the supplier are critical. Even having done all that, though, the company will still need to be assured of the quality of incoming purchased material.

One way to gain this assurance is via source inspection; in other words, have all suppliers certified on quality and do a complete job of inspecting their output. Until a company has achieved this status with all of its suppliers, however, it will need formal and objective measurements of the quality level of purchased items coming in the door.

Traditionally, most companies measure quality based on lots accepted and lots rejected. The problem with this measurement is that it does not give the buyer the information needed to eliminate defects at the supplier's. The way a "lots accepted, lots rejected" measurement works is a sample of the supplier shipment is taken when it is received and measured against the specification. Tolerances are established on each measurement and a small amount of variant material is considered acceptable.

For example, if a particular test is measured to a 1.0 acceptable quality level (AQL), the shipment will be accepted if it has five or fewer defects in a sample of 125 parts, rejected if it has six or more defects. If it is measured against a 0.5 AQL, it will be accepted if it has three or fewer defects. In other words, the more critical the test, the fewer the defects allowed in the sample. A supplier shipment is received and contains only three defects in the sample, so it is accepted and recorded as a good lot. In a lots accepted, lots rejected measurement system, the buyer does not see the three defects, only the rating of accepted. Even though the shipment is acceptable, the buyer must see the defects so a program can be developed with the supplier to eliminate those defects in all future shipments. The goal is zero defects.

A good quality measurement system that records defects would work as follows. When a shipment is received, Quality Control would pull and inspect a sample of that shipment. Based on this inspection, a lot will either be accepted or rejected. The quality measurement would include a record of the number of defective parts in the sample taken and the type of defect, whether or not the shipment was rejected.

For example, suppose an inspector pulls one hundred samples from an incoming shipment of one thousand parts and three are found to be defective for various reasons. In this case, the shipment is 97 percent in-specification. Every shipment that arrives on the receiving dock would be inspected in this way. The results are recorded against the particular supplier who produced the shipment and summarized weekly or monthly. (See Figure 12.6.)

This report then becomes the basis for the continuous process-improvement effort in the quality area. The report has identified the reasons for defects (grease, rust, color, etc.). If, in working with your suppliers, you are able to determine the cause of the defect (machine setup, tool quality, operator training, etc.), in time you should be able to

ABC Company
Month of July

	Samples Pulled	Defective	%	Reason
Part #127	100	3	97	Grease
Part #213	150	3	98	Rust
Part #543	50	3	94	Color
Monthly Total	300	9	97%	

Figure 12.6

eliminate the cause of the defect. By eliminating the cause, you should expect overall performance to improve. In this way, companies can identify the parts and suppliers that are consistently a problem.

If the current supplier is unable on his own to resolve the problem, you have the option of either sending your quality and engineering people to his manufacturing facility to help resolve the problem, or finding alternative suppliers who can meet your quality objectives. The goal is to initially have a minimum of 98 percent of all items within specification to gain control over your purchasing environment. The ultimate goal is 100 percent.

Again, you do not need an on-line program to measure quality. As in the case of delivery, you can initially use a personal computer to gather the data. It would be difficult to manually measure quality because of the high level of detail numbers involved in the computations. A one-part form with the information shown in Figure 12.7 should be sufficient for collecting the appropriate data.

One last point needs to be made: Rejected shipments should not count as on-time shipments. A rejected lot means that the supplier receives a poor performance rating not only on quality but also on delivery. Only good parts are considered on-time.

Price
Few companies measure price performance for each supplier and buyer to find out who is doing a good job of controlling costs and who is not. While many buyers can boast of their cost-reduction programs, few can

---

QUALITY

DATA INSPECTED _____

INSPECTED BY _____

ECN NUMBER _____

QUANTITY INSPECTED _____

QUANTITY REJECTED _____

REJECT NUMBER _____

TOTAL QUANTITY DEFECTIVE _____

DEFECT CODE _____

DEFECT DESCRIPTION _____

Figure 12.7

talk about the total impact of all price increases and decreases on the bottom line of the profit and loss statement.

Is pricing, delivery, or quality a company's most important measurement of performance? For many companies, delivery and quality are probably more critical than price. It is far better to purchase in uneconomical quantities and keep the shop running than to get a "real buy" and shut the line down waiting for delivery. But price is still an important factor if a company is to remain competitive, so it is vital to measure performance in all three areas.

Many companies measure price against a standard cost. In these cases, the price measurement is actually more of a reflection of how well the standard was established than of how wisely the buyers are using the company's financial resources. Suppose that on January 1 a supplier's price is $1. Purchasing estimates an 8 percent inflation factor, so the estimated cost at the end of the year would be $1.08. If the standard cost was calculated as the average cost for the year, Purchasing would place the standard cost at $1.04. If, however, Purchasing guessed wrong and the inflation factor was really 10 percent, the buyer's price performance would likely show an unfavorable variance of 2 percent at the end of the year. The buyer has little or no control over this factor.

Companies should measure actual performance rather than measure against a standard cost; it's much fairer to the buyers to measure them in a way that takes into account all the price increases and decreases and gives a weighted price index of performance. That price index, calculated for each item, is shown in Figure 12.8.

Each item would be calculated the same way and the total would be summed by the buyer or the supplier. (See Figure 12.9.) Using this technique, Purchasing has an objective measure of price performance for each buyer and supplier. This makes it possible to determine that Supplier A (80 percent on-time, 80 percent within specification, and 112

Price at end of year	$1.06
Price at start of year	1.00
Difference	.06
Annual Usage	10,000
Annual Price Impact	+$600.00

Figure 12.8

Part #	Price EOY	Price SOY	Difference	Annual Quantity	Annual $	Price Impact
137	1.06	1.00	.06	10,000	$10,000	+$600
243	.47	.45	.02	100,000	45,000	+2000
549	.03	.03	—	10,000	300	—
637	.63	.67	.04	6,000	4,020	−240
					$59,320	+$2360
					Price Index	104

Figure 12.9

percent on the price index) is not as good as Supplier B (who is 98 percent on-time, 97 percent within specification, and 103 percent on the price index). The same principle is equally applicable to buyers.

Lead Times
As we discussed earlier, supplier scheduling allows companies to reduce their lead times significantly. Basically, supplier scheduling represents a change from "paper" lead times to true manufacturing lead times. As the suppliers learn to use the information available to them in the supplier schedule to do a better job of planning raw materials and capacity, they will feel more comfortable working to manufacturing lead times. Purchasing should measure lead times regularly to see how well their suppliers understand the system. If the desired results aren't being achieved, more education may be necessary for the supplier or buyer involved.

Purchasing can measure two areas regarding lead times. First, of all the items purchased, how many have shorter lead times than before MRP II implementation? Once supplier scheduling is working, this answer should approach 100 percent. The second measurement would be the percentage of reduction in lead times. This would vary from company to company, depending upon the commodities purchased, but the result should be fairly substantial in most companies. Companies that work with their suppliers in a continuous improvement program in which they (1) assist their suppliers in cutting lead times by showing them how to set up cellular manufacturing and (2) reduce tool setup times yield lead-time reductions of up to 90 percent.

Inventory Investment
Since Purchasing is able to bring in the items to true need dates, rather than invalid due dates, substantial inventory reductions are possible. The Class A/B users of MRP II surveyed reported they typically saw a 30 percent reduction in purchased material inventory levels—in the face of inflation and, in many of the companies, high growth. (See Appendix B.) Double-digit inventory turns (12 to 18) are very common in leading MRP II/JIT companies today.

Schedule Completions in the Plant
One key test of the success of any MRP II implementation is customer service—the number of orders being shipped complete and on time. That number should be measured weekly to ensure that all departments

are executing the schedule correctly. The plant is a "customer" of Purchasing, so measuring Purchasing's performance amounts to measuring the "customer service" that Purchasing is providing to the plant. Similarly, the Sales Department is the "customer" of the plant, and the plant should be measured accordingly; its "customer service" measure is performance to the master production schedule.

Some of the best Class A MRP II companies publish weekly their schedule completions by product line. If a date is missed, it is recorded, as is the reason for the delay. This leads to identifying the root cause of the problem and solving it. Things that get measured get better. Do it right the first time.

Cost Reduction/Value Analysis

Purchasing should establish cost reduction/value analysis goals for each buyer and measure the buyer against those goals. MRP II helps the buyer in this area by identifying the items with the biggest annual savings potential. Purchasing can extract the annual usage of each item from the MRP II and dollarize it. Then, by printing the list of all the buyer's items in descending dollar sequence, the buyer can concentrate on those items with the greatest potential payback. Purchasing should measure the dollars saved against the goals set and routinely publish the results.

Inbound Freight Cost Reduction

In many companies today, inbound freight costs are a substantial part of the cost of materials. These costs can run up to 15 percent of the total dollars spent in Purchasing, although they are seldom broken out separately and typically are not visible to buyers. Because of the lack of visibility of these costs and the amount of money spent on inbound freight, there is a large potential for cost reduction in this area. You should see savings in this area when you implement an MRP II system for several reasons.

First, the buyers should have time to negotiate freight rates and study which mode of transportation is the most economical for a particular item. In the past this wasn't the case, and it was often left up to the supplier which carrier to choose.

Second, if a Traffic Department was assigned the task of specifying carriers or modes of transportation, or the buyer had time to do it, it was normally based on past experience. Traffic or the buyer would take bills of lading from previous shipments, and negotiated rates were based on what you did in the past. With supplier scheduling, you can give carriers future visibility of inbound freight by area and negotiate rates based on what is going to happen in the future. The way to do this is to convert scheduled receipts and planned orders into weight values by area. Figure 12.10 demonstrates how this works. A pooled freight rate based on a 15,800-pound shipment would be negotiated and all three suppliers would be instructed to arrange for pickup of their material on the same day of the week. The carrier would move all the material as if it were one shipment. As a result, your freight rate should be lower than it would be if these materials moved as five separate individual smaller shipments.

The third reason for reduced freight is the reduced need for premium air shipments due to expediting shortages. With valid schedules, more time to plan, better on-time supplier performance and fewer quality rejections, the need for premium freight should be minimal. Documented savings of up to 40 percent of the total freight costs are not

WESTERN MICHIGAN SUPPLIER GROUP

Supplier A Freight Plan in Pounds

	w/o July 15	w/o July 22	w/o July 27
Part 123	450	225	450
Part 456	730	0	730
	1180	225	1180

Supplier B Freight Plan in Pounds

Part 246	4500	7000	4500
Part 789	2000	2000	2000
	6500	9000	6500

Supplier C Freight Plan in Pounds

Part 137	7000	5000	7000
Part 973	1120	1150	1180
	8120	6150	8180

Summary—Total Freight Region

15,800	15,375	15,860

Figure 12.10

uncommon in good MRP II/JIT companies. (See Chapter 8 for a discussion on freight cost-reduction techniques.)

Intangible Results

Purchasing can derive intangible benefits from MRP II. First, there is improved credibility. When the supplier scheduler tells the supplier that an item is needed on a certain date, the supplier believes it because he understands the MRP II system. When Purchasing tells the plant that an item will arrive at 8:00 A.M. on Wednesday, plant personnel believe it, and on Tuesday night they set up the equipment to run the part.

There is also a significant improvement in quality of life. Purchasing has valid data to work with and knows what is expected. Buyers have retired the old "fire truck" and are out in front, managing their commodities. If there is a problem, the suppliers are informing the company in advance so there is time to work out alternative solutions. Everyone is part of the same team and recognized for his or her contributions.

Finally, the Purchasing Department becomes more motivated and professional. With time to perform the really important parts of their jobs, buyers are more valuable to the company. They save the company money because they are negotiating contracts and tracking costs instead of putting out fires. Using the future visibility provided by MRP II, the buyer can make better decisions and take actions that will have very positive long-term effects.

Continuous Improvement

The goal of any company should be to continuously improve their performance. No matter where they are today, they should seek to be better tomorrow. Purchasing should be a part of that drive to continuously improve. The problem is, many companies don't know where to start or what the ultimate goals should be. The Oliver Wight Companies have developed a series of questions, the ABCD checklist, to guide a company toward excellence. In Appendix C, we have listed those questions to guide you in your quest toward excellence.

SUMMARY

- Performance measurements are the only way to gauge your and your supplier's effectiveness in today's purchasing environment.
- Performance measurements must be cast in terms of concrete,

quantitative results and goals—subjective and generalized feedback cannot lead to improvement.

- Performance measurements are not designed to be used as a means of punishment or retribution; rather they provide feedback for buyers and suppliers to use to improve their performance.

- Measure performance in incremental steps against "benchmarks," interim goals that will gradually move a supplier to the desired performance level at a realistic pace.

- A comprehensive performance-measurement system covers delivery, quality, price, lead times, inventory investment, schedule completions in the plant, cost reduction/value analysis, and inbound freight cost reduction.

- Supplier delivery performance must be gauged on need dates, not replenishment dates; otherwise, "on-time" becomes meaningless.

- A 2 percent defect rate may, on the surface, seem like good performance. But two percent means twenty thousand defective parts out of every million, which is clearly unacceptable. Strive for 100 percent.

- Price should not be regarded as secondary to quality and delivery. An objective measure of price performance looks at *actual performance* rather than at performance against a standard cost.

- Lead-time reductions should be measured pre- and post-MRP II (all lead times should be shorter once supplier scheduling is in place). They should also be rated on an overall percentage basis.

- On-hand inventories of purchased parts should drop when items are delivered to true need dates.

- Scheduled completion dates are ultimately reflected in customer service, which should be assessed on a regular basis.

- Purchasing should develop a continuous improvement program and strive for purchasing excellence.

Chapter Thirteen

MRP II and JIT/TQC in Purchasing— Making It Work

Developing excellent suppliers to support MRP II and JIT/TQC means more than scheduling a one-time Supplier Education Day or mailing out instructions to the suppliers on their role in supporting your program. Rather, your efforts should be directed at assisting your suppliers in becoming excellent users of the same tools you rely on—MRP II and JIT/TQC—and to integrate their expertise with yours.

Unfortunately, when these tools first became available, many companies began to reduce purchase inventory, reduce purchase lot sizes, and increase frequency of supplier deliveries, while at the same time expecting their suppliers to ship defect-free product without first implementing the tools at their own facilities. These companies did not provide the suppliers with valid schedules, nor did they level production schedules, internally reduce lot sizes or lead times, stabilize product design, or implement a quality program. Not surprisingly, since the suppliers ended up carrying the inventory and all the extra costs, many of the early programs failed. And to add salt to the wounds, many of the suppliers involved became cynical about MRP II and JIT/TQC programs.

To avoid repeating such mistakes, companies must implement the programs internally first, find out how to correctly use MRP II and JIT/TQC, and then show their suppliers how to do it. This way, the suppliers will see the results at the buyer's company, understand that it is not just a program to push costs onto their ledgers, and be more willing to sign up to be involved. When the supplier sees how the buying company has developed valid schedules, improved internal quality, reduced lead times and costs, he will be more willing to support the program. Then it is time to start implementing supplier scheduling in purchasing.

A MAJOR CHALLENGE

When you implement supplier scheduling, whether you are going to use a supplier schedule or a Kanban mechanism and JIT/TQC philosophies, the effort must be integrated within your company's MRP II and JIT/TQC framework. As we just explained, these items need to be in place so you can show your suppliers how they work. Implementing supplier scheduling is not something one takes lightly; it is a major undertaking that must be carefully planned and executed if it is to succeed.

A detailed implementation plan is listed below to help define the key tasks that need to be covered by the purchasing spinoff task force (the numbers on the right side represent months, as in first month, second month, etc.). You will need to insert any additional steps that are needed at your company.

PURCHASING DETAILED IMPLEMENTATION PLAN

TASK	DATE
1. Outside Education	
A. Purchasing manager—MRP II and JIT/TQC class as a part of first-cut education	1
B. Other key purchasing people—MRP II and JIT/TQC class	3
C. Purchasing manager, other key people—class on MRP II in purchasing	3
D. Supervisor of supplier scheduling—MRP II class	4
E. Supervisor of supplier scheduling—MRP II in purchasing class	5
2. Inside Education	
A. Purchasing manager, along with all other department managers—a series of business meetings, over an eight-week period, based on an internal video-based education plan	3–4
B. All purchasing staff, buyers, and supplier scheduling supervisor—internal education, a series of business meetings, with purchasing manager as discussion leader	5–8

TASK   **DATE**

    C. Supplier schedulers—internal education, a series of business meetings, with supplier scheduling supervisor as discussion leader   9–12

3. Specify the Tools
    A. 1. Design the supplier schedule   9
          2. Supplier schedule due from systems group   10
    B. 1. Design measurement reports—delivery, quality, price   9
          2. Reports due from systems group   12
    C. 1. Design management reports—negotiation report, purchase commitment report, other supporting reports   9
          2. Reports due from systems group   14
    D. Design Kanban mechanisms, JIT program   12
    E. Design TQC program   12

4. Organization Defined
    A. Supervisor of supplier scheduling selected   4
    B. Reporting structure defined—department to which supplier scheduling group reports established   5
    C. Job descriptions developed for buyers and supplier schedulers   6
    D. Supplier schedulers selected   8

5. Supplier Education
    A. Define if one session or three sessions (MRP II, JIT, and TQC)   9
    B. Agenda for supplier education   10
        1. Teaching materials selected
        2. Supplier education manual developed
        3. Supplier commitment and action plan developed
    C. Pilot supplier selected   10
    D. Pilot supplier education with key supplier   11
    E. Education of the 20 percent of the suppliers who represent 80 percent of the dollars   12
    F. Education of the balance of production item suppliers   13–14
    G. Education of the suppliers supplying MRO items (maintenance, repair, operating supplies)   20–23

6. Pilot and Cutover
    A. Pilot with key supplier and fine-tune the system   12
    B. Cut over 20 percent of the suppliers who represent 80 percent of the volume (provided pilot is satisfactory)   13
    C. Measure results
        1. Pilot measurement reports   12–13
        2. Measure results—pilot and 20 percent of the suppliers   13–14
    D. Cut over remaining suppliers—production items only   15–18
    E. Measure results for all suppliers   16–18
        1. Follow-up education—specific suppliers   18
        2. Goals and action plans for all suppliers established and signed off by suppliers   18

PURCHASING DETAILED IMPLEMENTATION PLAN (*cont'd.*)

TASK	DATE
7. MRO Items Added to MRP II	
A. Plan established	19
1. Operating supplies	
2. Preventive maintenance items	
3. Balance of the items	
B. Operating supplies added to bill of material	20
C. Supplier scheduling of operating supplies	21
D. Preventive maintenance bill of materials developed	
and master production schedule established	22
E. Supplier scheduling of preventive maintenance parts	23
F. Balance of items—office supplies, repair parts, etc.	23–24
1. Establish item numbers	23
2. Establish inventory location and on-hand balances	23
3. Establish desired on-hand inventory and forecasted	
usage for each item and load to MRP	23
4. Supplier schedule balance of items	24

The purchasing manager must be a key part of the overall implementation effort. He or she must attend, as part of first-cut education, an in-depth class on MRP II and JIT/TQC. With this background, the purchasing manager can participate fully in the cost-justification process and should be able to make an informed commitment to the successful implementation of MRP II and JIT/TQC in purchasing.

In addition, the purchasing manager and perhaps other key people will require more specific education on how to design, implement, and operate an MRP system in purchasing. Also, the person selected to be the supervisor of the supplier scheduling group should attend both the MRP II and JIT/TQC class and the purchasing class.

The two educational experiences enable the attendees to develop a complete and valid implementation plan for MRP II and JIT/TQC in purchasing, and to assign responsibilities for the tasks in that plan. This gives them an opportunity to design the supplier schedule, the measurement reports, Kanban, and the other tools prior to beginning inside education for the rest of the department. With these in place in advance, the impact of the inside education process will be greatly strengthened.

Video-based education should begin in Month 3* and continue

---

* Throughout this chapter, we will make references to "Month 3," "Month 5," etc. These are not intended to be hard and fast times, but rather to represent approximately when various activities should begin and end.

through Month 8. The purchasing manager, as a member of the project team, would take part in the "teacher's course" and then lead the education of the entire department. This would include all purchasing personnel, as well as the supervisor of the supplier scheduling group, and would follow an education plan tailored specifically for the Purchasing Department in that company. The purchasing manager, possibly assisted by the other key purchasing personnel who attended outside classes, would lead the discussions on how supplier scheduling will be done at this particular company.

Regardless of which department the supplier schedulers will report to (Purchasing or elsewhere), they will also require specific purchasing education. If these people were formerly planners in Production and Inventory Control, they probably would have already attended the internal education for Production and Inventory Control (P&IC). If they are new hires or have been transferred from an unrelated department, they should receive their P&IC education along with specific purchasing and supplier scheduling education. These sessions should be led by the supervisor of supplier scheduling.

Next, Purchasing has to specify the informational tools necessary to do the job. These include: the supplier schedule, the measurement reports, and the negotiation and management reports. These tools should be initially defined before and refined during the education process, and their design should be finalized by around Month 9.

As of this writing, only a few MRP II software packages contain supplier scheduling capability. The odds are very high that a company implementing supplier scheduling will have to design and program its supplier schedule and possibly create other reports.

On the positive side, though, many MRP II software packages do contain the ability to maintain purchasing and supplier-related data. The programming task, therefore, becomes primarily one of "retrieval and display" programming, which is typically less difficult than starting from scratch.

THE SUPPLIER SCHEDULE

The supplier schedule itself is the first tool that needs to be discussed. Purchasing will have to answer questions such as: What will the format look like? How many weeks are going to be displayed to the suppliers? Will capacity units of measure for suppliers be included? Will the report

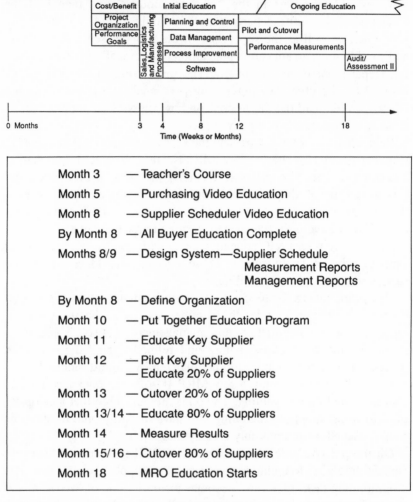

Figure 13.1

display the information horizontally (from the left side to the right side of the supplier schedule) or vertically (top to bottom)? Chapter 3 showed a horizontal format for a supplier schedule. Similar information, but in a vertical format, is shown in Figure 13.2.

The advantage of the vertical format is that it can display an unlimited number of weeks into the future, whereas a horizontal format is somewhat more limited in the number of weeks shown. A potential disadvantage of the vertical format is that it may reduce your ability to quickly

JONES COMPANY SUPPLIER SCHEDULE FOR: SMITH, INC. WEEK OF 02/01			
Supplier Scheduler: AB **Buyer: CD**			
	Order Status	*Date Due*	*Quantity*
PART NUMBER 13579 PLATE	Scheduled	02/01	100
LEAD TIME 4 WEEKS	Scheduled	02/22	100
	Planned	03/08	100
	Planned	05/10	100
	Planned	07/12	100
PART NUMBER 24680 PANEL	Scheduled	03/01	50
LEAD TIME 6 WEEKS	Planned	04/19	50
	Planned	06/28	50

Figure 13.2

see all the items due from a supplier in a given week. Purchasing must make these decisions early enough to allow the systems and data processing people time to program, test, and debug the supplier schedule.

SUPPLIER MEASUREMENT REPORTS

The second tool involves supplier measurement reports (discussed in detail in Chapter 12). Purchasing must answer questions such as: How is the company going to measure delivery, quality, and price? How often does Purchasing want the measurement reports and what should they look like? Are both suppliers and buyers going to be measured? How about the supplier schedulers? How long should data be kept? Which of this information should be accessible by CRT?

Finally, Purchasing needs management tools for use once MRP II and supplier scheduling are "up and running." These can include negotiation reports, purchase commitment reports, and perhaps other supporting information to allow purchasing people to be effective negotiators and buyers.

On the sample negotiation report shown in Figure 13.3, all the items in the commodity group are displayed in a descending dollar sequence. It shows the annual usage in dollars, the average weekly usage, the safety stock and lot size information, the current on-hand inventory, and the total dollars used annually in that commodity. This gives the buyer good information to allow negotiation of annual contracts and effective

Negotiation report by commodity code
commodity code 01 and buyer code 80

Part	Description	Safety Stock Wks.	Qty.	Lot Size Wks.	Qty.	On-Hand Invent. Wks.	Amt.	Unit Cost	Annual Usage	Avg. Wkly. Usage
896890009	FIBERGLAS	1	75	1	75	1	$767	$ 7.67	$32,704	82
896890008	FIBERGLAS		5		15		$ 0	$13.50	$16,848	24
896890006	FIBERGLAS	5	25	5	25	7	$135	$ 3.87	$ 1,006	5
						TOTAL ————	$902	—————	$50,558	

Figure 13.3

management of that commodity. This negotiation information could also be shown by supplier, plant, etc.

Figure 13.4 shows a typical purchasing commitment report. This report takes the scheduled receipts (shown as an *) and the planned orders (shown as a P) and translates those orders into financial terms in the week they are due from the supplier. The right-hand column indicates to the buyer the total dollars committed to each supplier over the next thirteen weeks. The bottom total indicates the total dollars committed to all suppliers in a given week. This aids the buyer in managing supplier commitments, future inventory dollars, and the company's weekly requirement for cash flow to support purchases.

Many other kinds of information are also available. The challenge, in terms of implementation, is to define specifically what's needed in reports, and with what frequency the information must be presented.

DEFINING THE ORGANIZATION

A clear definition of the supplier scheduling organization is also necessary for the implementation to succeed. Following the selection of the supplier scheduling supervisor, it is necessary to establish the reporting structure of the supplier scheduling group. As we state in Chapter 5, it is preferable to have this group report to Purchasing, all things being equal. But all things are not always equal; existing organizational structures, geography, the personalities and preferences of people, and other factors may play a part in this decision.

Following this decision, job descriptions for the buyers and supplier schedulers should be developed. The responsibilities of both the supplier scheduler and buyer, as discussed in several chapters, should be clearly defined in the individual job descriptions. Finally, the supplier schedulers should be selected by Month 8 so that their purchasing education and training can start the following month.

SUPPLIER EDUCATION

MRP II, JIT, and TQC education cannot be limited only to the people employed by the company. Suppliers are key to the success of each of these programs, and therefore they need to understand the disciplines involved in education and training.

The typical company would have three different education sessions

PURCHASING COMMITMENT
BY VENDOR
FOR WEEK ENDING 01/21/XX

(*) SCHEDULED RECEIPTS
(P) PLANNED ORDERS

DATE 01/19/XX

VENDOR		CURRENT & PAST DUE	1/28	2/04	2/11	2/18	2/25	NEXT 4 WEEKS	FOLLOWING 4 WEEKS	VENDOR TOTAL
SMITH COIL EZ	*	2,975.50	743.00	100.50	1,772.50	525.00				6,116.50
	P						575.00	6,683.00	3,199.75	10,457.75
WILSON PLASTICS FA	*	388.36	1,431.10							1,819.46
	P			907.10	908.10	1,254.27	461.21	4,655.20	3,787.37	11,973.25
P B PECK FB	*	4,880.28	6,002.00	11,851.60	10,087.80	5,281.55				38,103.23
	P			1,467.30	1,164.68	1,849.74	8,052.28	44,064.24	40,538.33	97,136.57
HARTMAN FD	*	2,193.01								2,193.01
	P							1,737.60	1,685.60	3,423.20
TODSON FE	*	5,442.90	6,550.00	4,987.50	135.05	430.00				17,545.45
	P				5,435.95	4,988.35	5,675.40	29,272.01	36,917.51	82,289.22
JAMES FF	*	3,838.85	4,215.20	1,951.00	2,731.00	1,988.40				14,724.45
	P						925.00	5,687.00	8,172.74	14,784.74
TOTAL ALL VENDORS		19,718.90	18,941.30	21,265.00	22,235.08	16,317.31	15,688.89	92,099.05	94,301.30	300,566.83

Figure 13.4

for their suppliers. First, they would educate their suppliers on MRP II and work to get valid schedules and supplier performance to the schedules. Second, they would educate their suppliers on JIT, eliminating any waste in the supply pipeline. Lastly, they would do the TQC education on defect-free product delivered to the point-of-use in the shop.

You should consider bringing suppliers into the plant for a one-day supplier education program to kick off each of the programs whenever practical. In some cases, due perhaps to geography or other factors, it may be necessary to "take the show on the road"—that is, to conduct the education day at the supplier's plant. This approach can work well, although bringing the suppliers into one's plant and showing them how it works seems to have more overall impact and generate better results. It is important to have several people present from a given supplier at the training, and the mix of people in attendance would vary by topic.

For the MRP II education, supplier attendees should include their customer service/order entry manager, the manager of the scheduling group, and an executive who can commit to working from schedules rather than purchase orders, in addition to the local salesman.

For the JIT education, the group should include Engineering and Manufacturing, and the TQC education should include the quality group from the suppliers. Supplier education is a very time-consuming effort. If you have four hundred suppliers and you educate twenty at a time, you will have at least twenty sessions on each topic. If you do one supplier at a time, you will never finish. The expectation and goal is that you will educate all suppliers on MRP II, JIT, and TQC.

The supplier education process should consist of basic education on each of these areas, plus more specific training on "how to" in each area. In MRP II, this would cover how the supplier scheduling process works, what is expected of a supplier under a supplier scheduling concept, and how the partnership arrangement will work. In JIT, this would cover how the Kanban concept works, how the supplier schedule and Kanban relate to each other, how waste can be eliminated in the system, and what is expected of a JIT supplier. In TQC, this would cover basic TQC principles, the concept of quality at the source, the ideal of defect-free product delivered right to the point-of-use, and what is expected of a TQC supplier. Each supplier education session would have a different purpose and a different agenda. It is hard to envision how a company could combine these topics and do justice to any one of the areas covered. The agendas for the Supplier Education

Day on MRP II should be developed by around Month 10, as should outlines be developed for what will be covered in the JIT and TQC Supplier Education Days.

It is also necessary to develop a supplier education manual for each of these topics. This is a document that the suppliers can take back to their companies for use in educating their own employees. It would cover the same topics as were covered during the Supplier Education Day, although greater detail may be necessary to aid in educating people at the supplier's plant.

Below is a sample agenda for a Supplier Education Day. Obviously, each company will vary somewhat from this agenda depending upon the size of the company, the site of the education, the level of expertise in the company, etc. It is only a guide to demonstrate what some companies are doing today.

Supplier Open House
Sample Agenda

8:00– 8:30	Registration
8:30– 8:45	Welcome
8:45– 9:45	Company vision
9:45–10:00	Break
10:00–12:00	MRP II or JIT or TQC education
12:00– 1:00	Lunch
1:00– 3:00	Factory tour
3:00– 3:15	Break
3:15– 4:15	Expectations and supplier schedules
4:15– 4:45	Panel question-and-answer period
4:45– 5:00	Commitment from suppliers and request for action plan from suppliers

An effective Supplier Education Day might be structured in the following way. (We will use an MRP II example, but the same basic structure would apply if it were a JIT or TQC Education Day.) After the welcome of attendees, a top executive of the company would deliver a company vision presentation. This would outline the company orientation and its strategies, define why the MRP II program was started, what is taking place in the marketplace to cause the need for MRP II, what the competition is doing in this area, an overview of MRP II and what it will accomplish for the company, why the attendee has been asked to participate in the session, and how top management envisions the company's future relationship with suppliers.

After a short coffee break, the next two hours would be devoted to general MRP II education, based on some of the same material (videotape plus training aids) used to educate the buyers and supplier schedulers. This session should cover the following topics: fundamentals and applications of MRP II, the mechanics of executing the purchasing plan, and how to operate MRP II in purchasing.

After lunch, a factory tour would be a good idea. This would be the buyer's chance to show the suppliers that the company is serious about MRP II and is actually using it, not just talking about it. If done correctly, this will show the suppliers that they are not participating in "just another" program, but that MRP II is a way of life for your entire company. If this Education Day is not held at the factory, then a slide show of what you are actually doing, led by a supplier scheduler, master scheduler, and manufacturing supervisor, can often achieve good results.

Following the afternoon break, the emphasis would shift to the expectations for the future and the supplier schedule. The discussion should cover the format of the supplier schedule, what suppliers are authorized to produce (scheduled receipts), and how they can plan raw material and capacity, capacity units of measure, etc. This would also be a good time to verify lead times, lot sizes, minimum order quantity, maximum weekly supplier capacities on each item number, supplier plans for vacation shutdowns, etc. Time should be set aside to discuss the day of the week that the supplier schedule would be sent to the supplier, cutoff dates for the previous week's receipts, and the need for feedback on past-due orders.

The principle of "Silence is approval" should also be stressed during this time. Suppliers must understand that they will be held accountable for delivery on the dates shown on the supplier schedule, and it is imperative that they communicate in advance if they will miss a delivery date.

A quick review of the supplier manual should also be done during the after-break session. Since this is the document that suppliers will use to educate their people back home, they must have an overview of what is in it. The manual should cover MRP, how the supplier schedule works, and what is required of suppliers.

Some time should also be provided for questions from the suppliers.

Questions are best addressed by a panel consisting of the day's presenters, as well as the top company executive who described the company vision.

The Supplier Education Day should be completed with a commitment from the supplier to work to the supplier schedule. After that commitment, the buyer and supplier need to work together to develop an action plan to bring performance on delivery and quality up to acceptable levels. Suppliers must realize that they will now be receiving valid need dates, and therefore that they must ship high-quality parts on time consistently. The action plan should spell out the supplier's current performance levels on delivery and quality. A timetable should be developed that shows when the supplier would be at acceptable levels of performance in the areas of delivery and quality to support the supplier scheduling program.

After designing the supplier education program, it is time to test it with a pilot (see below). This first "live run" of supplier education should occur by Month 11. Month 12 would be spent educating the 20 percent of the suppliers who represent 80 percent of the dollars spent on purchased items.

PILOT AND CUTOVER

With the training complete, it is time to implement supplier scheduling, JIT, or TQC. Its usually best to use a pilot with a selected supplier to be sure the program is working correctly before cutting over a large number of suppliers.

The best choice for a pilot is a key supplier who is fairly close geographically, well organized, well run, and willing to work with you to develop and perfect your supplier education program for supplier scheduling, JIT, or TQC. This key supplier should be selected by Month 10 and, as we said earlier, educated by around Month 11. This will give Purchasing a chance to get feedback on the first Supplier Education Day from the pilot supplier and make changes before educating other suppliers.

The actual pilot with the key supplier starts around Month 12 and normally lasts for three or four weeks (longer if the pilot is not working properly—in that case, stop and correct the problem before going any further). Based on the feedback from the pilot supplier, fine-tuning of the supplier scheduling system, the JIT Kanban pilot, or the

TQC program occurs during the pilot period. The measurement reports should also be piloted with the key supplier at the same time.

Bear in mind that suppliers will not automatically become perfect the minute they start to receive their supplier schedules or Kanban mechanism. Initially, they will still have quality and delivery problems. The last thing a company needs when cutting over onto supplier scheduling or JIT is to have serious stockouts on purchased items and to shut down production. Therefore, it is critical that Purchasing work closely with the suppliers during and after the pilot phase to quickly resolve any problems that do occur. In this way, the supplier scheduling concept has a good chance of succeeding.

It is often a good idea during pilot and cutover to make use of safety stock and/or safety time to protect against supplier problems. Then, as supplier performance improves, these "cushions" can be removed. Remember priority number one when implementing supplier scheduling: run the business.

The 20 percent of the suppliers who represent 80 percent of the dollars purchased should be cut over to supplier scheduling beginning around Month 13. During Months 13 and 14, you should be able to begin seeing the improved results of having the pilot supplier and the larger suppliers on the system. By this time, better on-time delivery, lower inventory, reduced lead times, etc., should start to appear. If these results are not occurring, something is wrong, either with the supplier schedule itself or the education program. Again, should this happen, go no further—fix what is wrong first.

When you begin to see some results, you can begin to cut over the 80 percent of the suppliers who represent the last 20 percent of the dollars purchased on production items. This cutover would occur around Month 15, with all production items ideally on supplier scheduling by Month 18.

During Months 16 through 18, companies should be measuring results with all suppliers. They should determine which suppliers are not improving their performance on delivery and quality during the first three months, and provide these suppliers with follow-up education. Goals for all of the suppliers should be established on delivery and quality, and action plans should be established to get them up to acceptable levels of performance. The supplier should be in agreement with these goals and action plans and commit to them in writing.

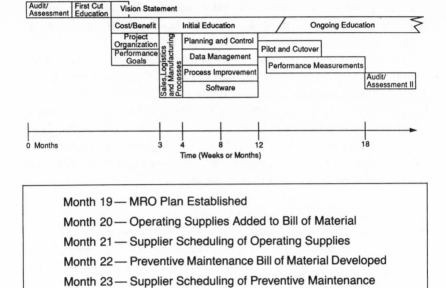

Month 19 — MRO Plan Established

Month 20 — Operating Supplies Added to Bill of Material

Month 21 — Supplier Scheduling of Operating Supplies

Month 22 — Preventive Maintenance Bill of Material Developed

Month 23 — Supplier Scheduling of Preventive Maintenance

Month 24 — Balance of MRO Items on Supplier Scheduling

Figure 13.5

MRO ITEMS

With all the production items on supplier scheduling by around Month 18, MRO items can be cut over during Months 19 through 24. (See Figure 13.5.)

The detailed plans to add MRO items to the supplier scheduling system should be established by Month 19. Operating supplies can be added to the bills of material by Month 20, with supplier scheduling of these supplies starting in Month 21. Bills of material for preventive maintenance items could be developed during Month 22, and the preventive maintenance schedule structured in such a way that it becomes, in effect, the master production schedule for maintenance. Supplier scheduling of preventive maintenance parts could then start in Month 23.

The balance of the items the company buys, such as office supplies and repair parts, could be added during Months 23 and 24. Part numbers would have to be established for all the remaining items, along with inventory locations. Inventory balances should be verified. Also during this time, it is necessary to establish the desired minimum on-hand inventory and forecasted usage for each item and load them into the system. Supplier scheduling of those items could then begin.

Given this schedule, by the end of Month 24, all production and MRO items would be supplier scheduled either with a supplier schedule, Kanban mechanism, or under a TQC program.

SUMMARY

- Without proper education, MRP II and JIT/TQC will not be as successful and will not generate their true benefits.

- Purchasing needs to define the formats of the supplier schedule, the measurement reports, and the management reports early enough to give systems time to program, test, and debug the programs.

- A clear definition of the supplier scheduling organization is necessary for the implementation to succeed.

- Suppliers are key to the success of MRP II and JIT/TQC, and therefore they need to understand each of these disciplines through education and training.

- It is advisable to develop a supplier education manual for each topic so that the suppliers have a document to take back home and use to educate their own employees.

- It is recommended that a pilot be used with a selected supplier to be sure the program is working correctly before cutting over a large number of suppliers.

Chapter Fourteen

Create a Spark, Ignite Change

As we said at the end of Chapter 1, consider working on MRP II and JIT immediately. If your company is not currently operating under MRP II or employing JIT, spearhead an effort to get them under way. Perhaps consolidate your supplier base, spreading your volume over fewer suppliers. Then start more frequent deliveries, maybe shifting from once a month to once a week. Once you have gotten accustomed to this way of doing business, you can begin working with your best suppliers to improve quality, eventually eliminating inspection of incoming materials. Continue gradually until you accomplish your full range of goals and achieve excellence at your company.

Just-in-Time means one less at a time. . . .

One less supplier. Pick an item and try to go from using three suppliers to two. Once that is complete, go from two to one. Then pick like items and consolidate them under one supplier. Reduce the supplier base as much as possible.

One less hour in placing purchase orders. Pick a supplier and use a supplier schedule to eliminate the need for purchase orders. Then pick another and another until all suppliers are on schedules.

One less piece of paper. With all suppliers on schedules, look to EDI or a Kanban mechanism to eliminate the need for the printed schedule. Pick a supplier and try EDI or Kanban, then another and another until all pieces of paper are eliminated.

One less in the safety stock. Pick a part that has a good history of supplier delivery performance and reduce the inventory. Work with that supplier in a joint program to get more frequent shipments and eventually, if possible, eliminate the need for any inventory on the part.

One less day in supplier lead time. Work with the supplier to understand the elements of lead time that make up the total lead time on the part. Pick one of those elements and try to reduce it a day, then another, until you get to the point of synchronized schedules.

One less supplier rejection. Pick an item that has a history of rejections and try to determine the underlying reasons. Eliminate one reason, then another, until there are no more rejections on that item. Then pick another item and another until all rejections are eliminated.

One less defect in quality. Once rejections have been eliminated, work to eliminate ALL defects. If you are at 1 percent defect level, 10,000 defects per million, try to lower that to 9,750, then to 9,500, and so on until all defects are eliminated and all suppliers and parts are certified for quality.

One less in the order quantity. Pick an item that has a fixed order quantity and ask why that quantity was selected. Challenge the reasons for the lot size (i.e., setup times, etc.) and reduce the order quantity as the reasons are resolved to match your internal lot sizes.

One less minute in setup at the supplier. Help your supplier understand the need for setup reductions. Then assist in taking a minute out of one setup, then another and another, until setup time is no longer an obstacle.

One less move on the shop floor. Pick an item, get the quality certified, then have the supplier deliver it to the point-of-use on the shop floor. Then do the same with another. The goal is to eliminate the need to move parts from Receiving to Quality Control to the warehouse and then to the shop floor.

Finally, one less at a time means seeking improvements over and over again in every area of purchasing and manufacturing, day after day, every day, until purchasing excellence is achieved.

Appendix A:

Questions for a Good Supplier

DELIVERY

You should expect your suppliers to answer "yes" to the following questions:

1. Do you measure on-time delivery performance to your customers? What is that percentage?

2. As part of a formal program, do you document the reasons you deliver late and have action plans in place to prevent the underlying problems from recurring?

3. Are you willing to work in a JIT environment where you are responsible to plan to my supplier schedules and deliver within a window of, say, -0, $+5$ days of my need dates?

4. Do you have a documented policy/procedure that states that you are to notify customers in advance if their requested delivery dates will be missed?

5. Would you be willing to give me the names of two or three companies you currently service so I can verify your service levels?

Ask your suppliers the following questions to help determine whether they can meet your desired lead times.

1. Are you willing to deliver daily or weekly in small quantities as specified on the supplier schedule?

2. Are you willing to discuss the elements of lead time (raw material acquisition time, scheduling time, and actual production time) and establish stable lead times to your actual production time?

3. Are you willing to plan raw materials and capacity based on my supplier schedule?

4. What are your current lead times?

Questions for the suppliers relating to packaging and delivery include:

1. What is your standard container size per part number?

2. Would you be willing to modify package size or the number of items in my containers if required?

3. Do you deliver in your own truck or use a common carrier?

4. If you deliver in your own truck, are you willing to use the same driver every time you deliver to me and drop the material off at the point-of-use on my manufacturing floor?

5. If you use a common carrier, are you willing to specify that the carrier must use the same driver every time the items are delivered to my facility, and that the driver delivers the items directly to the point-of-use?

Questions to ask in the area of flexibility include:

1. Do you have any flexibility in your manufacturing processes or are they fixed once you start to manufacture the item?

2. At what points in the process is the flexibility built in and what are the limits of change you can handle?

3. Given that both of us will have changes in our schedules due to the demands of the marketplace, are you willing to handle last-minute changes routinely, and is there a written policy that defines how changes are handled?

PRICE

Ask your suppliers the following questions regarding pricing:

1. I'm looking for a long-term, win-win relationship; are you willing to reveal the cost elements of the materials that I buy?

2. In the same vein, are you willing to reveal your desired profit margin on the items I buy?

3. Are you willing to allow my engineering staff to assist you in reducing the cost elements of your raw materials and manufacturing process?

4. Do you currently have an established, documented value analysis/ value engineering program in place, and what are the results of that program?

In terms of pricing for the long haul, ask the suppliers the following questions:

1. Are you willing to negotiate a price over time, say two years, based on my supplier schedule projections?

2. Are you willing to agree to a joint program to reduce the cost of the part over the life of the contract, with a lower price at the end of the contract period than at the beginning?

3. Are you willing to notify me in advance of all price increases, explaining what caused the increase, and work with me to either eliminate the need for the increase (i.e., alternative materials, etc.) or delay the increase?

QUALITY

Ask the following questions of suppliers to determine whether they are capable of giving you the quality and reliability you need:

1. Do you have a quality-assurance document guaranteeing me that material ready for packaging has passed final inspection and meets all of my requirements?

2. Do you have a quality-assurance document guaranteeing me that the material ready for packaging is the exact quantity I ordered?

3. Do you have a process for verifying that the shipping papers and the quantity to be shipped are the same?

4. Are special packaging and shipping requirements made available to and followed by the department personnel?

5. When a special marking, symbol, or bar-coded label is required, do you inspect it prior to packaging and shipping to ensure that it is correct?

6. Do your quality-assurance people periodically perform audits of the Packaging and Shipping Departments for conformance to approved procedures?

7. Will you guarantee that my material will not be damaged while in the packaging and shipping areas?

Typical questions to which suppliers should be able to say "yes" in the area of feedback include:

1. Do you conduct investigations to determine the cause of non-conforming materials?

2. Are corrective action plans initiated and the results of the corrective actions documented?

3. Do you routinely provide written feedback to your customers, documenting causes of problems, corrective actions, and the results of the quality-assurance program?

4. Is someone within your organization responsible for quality?

5. Do you document and file quality investigations in a manner that allows you to create an historical data base for problem solving?

Some questions to ask suppliers to ensure that they are building quality into the product include:

1. Do you conduct quality audits of your sources of supply to assure you that *your* suppliers will produce only acceptable quality raw materials?

2. Do you verify that your incoming raw materials conform to the specifications?

3. Do you maintain performance history records on supplies and materials?

4. Do your in-process inspections and/or tests assure conformance to specified requirements?

5. Are your tests and/or inspections performed before and during process adjustments, machine setup, tool changes, etc., to assure compliance to specifications?

6. Are there clearly written policies within your company stating that an operator is authorized to shut down the machine rather than make defective material?

7. Are control charts used to monitor your daily performance?

8. Do you initiate corrective action immediately when out-of-control or unstable processes are encountered?

Some questions to ask the suppliers regarding their process control are:

1. Do you perform process control planning on new products and on major changes to existing products?

2. Is quality planning a normal function of your quality management system for new products and for major changes in existing products?

3. Is the change control system governed by clearly written instructions and procedures?

4. Is it standard practice to notify all customers of changes to existing products in advance, allowing them time to evaluate the impact of the changes?

Here are some questions to ask suppliers regarding their attitude about the importance of quality:

1. Do you have an established quality organization?

2. Do you have clearly written directives that establish the responsibility, authority, and policy of your quality management system?

3. Have you established clearly defined quality goals and objectives?

4. Can you provide a case study that demonstrates your willingness to jointly resolve quality problems with your customers?

Questions you should ask suppliers regarding their full-service capabilities include:

1. Is quality planning a normal part of your quality management system for new products and for major changes in existing products?

2. Do you perform process control planning on new products and on major changes to existing products?

3. Is quality planning a normal function when you design or modify existing tooling, gauges, and equipment?

4. Can you provide a case study demonstrating your ability to take a concept, design and build tooling to actualize that concept, and then deliver quality items on time to the customer?

RESPONSIVENESS

Questions to ask your suppliers in this area of responsiveness are:

1. Who in Order Entry is assigned to my account? What is that person's telephone number?

2. Does that person understand how to read my supplier schedule and know what actions he or she is authorized to take based on the schedule?

3. Is there a customer service representative assigned to my account? What is that person's telephone number?

4. Is there a technical support person in my area assigned to my account? What is that person's telephone number?

5. Who is my quality contact if I have a problem, and what is that person's telephone number?

6. Can I have an organizational chart with telephone numbers of your company personnel?

LEAD TIME

In terms of lead times, you should be able to obtain the following information from your suppliers:

1. What are the elements of your lead time?
2. How do you schedule your suppliers and how much flexibility do they have in their lead times?
3. How long does it take to set up and actually run my items?
4. Are you willing to allow my engineers to assist you in reducing your internal setup and run times?

LOCATION

You should ask your suppliers the following questions regarding multiple locations:

1. Do you have more than one manufacturing location?
2. What is your capacity to make the items at each location?
3. Are all your plants equally automated?
4. Are any of the plants operating in a TQC/JIT mode today?
5. Do you have the ability to specify which location produces my material?
6. Who are your key subcontractors?
7. Do your subcontractors have TQC/JIT in place?
8. Can I talk to your subcontractors regarding quality, delivery, setup, etc.?
9. How is capacity allocated by customer in your facility?
10. What other products do you produce, and would you be interested in additional business in other product lines if there is a fit between our two companies?

TECHNICAL CAPABILITIES

Questions about technical capabilities include:

1. Are you a leader in new technology, or do you wait until a new technology proves itself?

2. What is your track record for bringing on new products and product improvements on time? Give me an example.

3. What is your manufacturing strategy for the next two years in terms of new products?

4. Are you constantly looking for ways to improve the current products you manufacture? Give me an example.

5. Is Total Quality a standard operating principle of your Research and Development group?

6. Can you give me an example of where your R&D Department successfully developed a new product for a customer?

R&D

You will want to determine your suppliers' development commitment with questions such as:

1. What is your R&D budget as a percentage of your sales?

2. How many people have you assigned to R&D?

3. If necessary, are you willing to allocate development money to my account?

4. Are you willing to work on joint development projects with your customers? Give me an example.

FINANCIAL AND BUSINESS STABILITY

In terms of financial and business stability, you should ask your suppliers questions such as:

1. Can you supply me with copies of your annual reports for the last two or three years?

2. Can you give me an indication of your cash flow, receivables, etc., to indicate your current financial stability?

3. Is there a high turnover in your management team or an expected change in ownership which could affect our long-term relationship?

Appendix B:

Purchasing Survey Results

BACKGROUND

A survey was taken of over one hundred Class A and B users of MRP II (see Appendix C for a definition of Class A, B, C, and D), focusing on the results they had obtained in the area of purchasing. The respondents were primarily purchasing professionals, MRP II project leaders, and production and inventory control managers. This appendix is a summary of the findings of that survey.

The businesses studied represent many different industries. The products they make include: floor sweepers, equipment for the telecommunications industry, cosmetics, pumps, diesel engines, hand tools, valves, forgings, pipe, wood kitchen cabinets, machine tools, instrumentation, trenchers, backhoes, power tools, office furniture, clothing, electronic instruments, cathode ray tubes, autoclaves, chemical processing equipment, photocopiers, ball bearings, fabricated parts, process chemicals, and both high-volume and low-volume parts.

These companies also represent many different types of manufacturing, including make-to-order, make-to-stock, assemble-to-order, and design-to-order. Many of the companies surveyed were leaders in their respective industries; all are located in North America.

The sizes of the companies surveyed varied widely. Their annual volume of purchases ranged from a low of $4 million to over $1 billion. Eighty percent of them purchased less than $30 million worth of goods annually.

RESULTS

Organization

Fifty-five percent of the companies surveyed have one or more persons in the role of supplier scheduler—in other words, someone other than the buyer who schedules the suppliers. Thirty percent of the companies have buyer/planners—one person who does the MRP planning, the supplier scheduling, and the buying. Fifteen percent still cling to the traditional "requisition" form of scheduling, in which a material planner writes out a requisition based upon the MRP output and the buyer converts it into a hard-copy purchase order.

Schedules

Of those respondents who did some form of supplier scheduling, 78 percent used a supplier schedule while 22 percent simply gave their computer-generated MRP output sheets directly to the suppliers.

Interestingly, several companies reported some form of computer-to-computer linkage (electronic data interchange) with their suppliers. One of those companies has placed CRTs in the offices of its key suppliers, allowing them to look directly into the company's MRP system at the requirements and orders.

Reporting Structure

The respondents were split neatly down the middle on the reporting structure of the supplier scheduling organization. In almost 50 percent of the companies, the supplier schedulers report to purchasing, and in the other nearly 50 percent, they report to production and inventory control. There was only one exception to the 50-50 split. In that company, the supplier scheduling group reports directly to the materials manager, who is on an equal level with managers of the purchasing department and production and inventory control.

Supplier Scheduler/Buyer Ratio

While the number of parts per buyer or per supplier scheduler varied widely from case to case, the average company surveyed reported one supplier scheduler as being able to support two buyers.

BENEFITS ACHIEVED

The benefits achieved in the Class A/B MRP II users' purchasing departments are impressive. The survey was designed to quantify benefits in four areas: on-time delivery, lead time reduction, inventory reduction, and purchase cost reduction. Respondents also commented on other areas of benefit, but these comments have not been quantified in any way.

On-Time Delivery

The keys to success in this area can be boiled down to a supplier schedule with valid need dates and intensive supplier education. On-time delivery was measured by comparing due dates with actual delivery dates. Companies using supplier schedules and supplier schedulers averaged an on-time supplier performance level of 97 percent. Those companies that used Kanban with supplier schedules averaged 99 percent on-time.

A few of the surveyed companies gave their MRP output directly to the suppliers. Those companies averaged 93 percent on-time service from their suppliers. We theorize that the 4 percent difference may be due to a small amount of misunderstanding by the suppliers' personnel on how to interpret the MRP reports.

Those companies that continue to use hard-copy purchase orders averaged on-time supplier delivery performance of only 76 percent. Apparently from the suppliers' viewpoint, where traditional purchase orders are used, there is not a great difference between their customers' old systems and MRP II. Therefore, supplier performance does not improve significantly until their customers change to supplier scheduling.

The typical respondent stated 30 to 40 percent of the open purchase orders were past due prior to MRP II implementation. That figure was reduced to between 15 and 20 percent after one month of MRP II because purchasing "rescheduled-out" all the past due orders that weren't needed yet. Five months later, through a combination of supplier education and valid schedules, the level of past due orders was reduced to 3 percent.

Another way to look at supplier delivery performance is in terms of items short in any one week. Users typically cut purchased item shortages from 30 to 50 parts in a week to 4 or 5. As one buyer stated: "We cut the anticipated delay report from two pages to less than half a page, and we know about all those shortages in advance from our suppliers." Many companies also commented that particular commodities which had been troublesome before were no longer problems.

Lead Time Reduction

Most companies surveyed saw a dramatic decrease in lead times. Many stated they had no lead times greater than eight weeks. One design-to-order company that is buying capacity from its sand casting supplier has cut the lead time from 26 weeks to eight weeks and eliminated shortages. They forecast capacity requirements to their suppliers at the eight-week window but show the actual configuration one week out, when the supplier is actually packing the sand in the mold box.

Inventory Reduction

The surveyed companies also reported significant inventory reductions made possible by MRP II. One company was able to cut its on-hand steel inventory from 13 weeks to five days, while simultaneously eliminating steel shortages. Another turned purchased item inventory 37 times a year, while a clothing manufacturer increased inventory turns on their purchased yarn from 50 to 125 turns. Responses of eight or more inventory turns per year on purchased items were very common. The average inventory reduction reported was 30 percent.

Purchase Cost Reduction

Companies with a supplier schedule and a supplier scheduler saw an average 13 percent annual reduction in the cost of purchased material. Those with a buyer/planner saw an average 7 percent annual cost reduction. We theorize that the difference in benefits is the result of the time the buyer/planner has to devote to material requirements planning and scheduling. Those companies still using hard-copy purchase orders saw a 2 percent annual cost reduction.

Comments

In general, the respondents were very positive about the use of MRP II and purchasing. Overwhelmingly, they reported they had better information and greater control over their departments than ever before. They knew exactly what was going to be built weekly on every product the company manufactured (through the master production schedule) and what the requirements were on every item they purchased (through MRP).

The only negative MRP II comments uncovered in the survey involved "nervousness" in the master production schedule. Several purchasing managers noted a stable master schedule is critical to purchasing and that their companies sometimes violated the established master schedule policies and time fences. However, all agreed that if the master schedule policies were followed, and all changes inside the firm time fences were reviewed with purchasing first, this nervousness would be minimized.

One purchasing manager summed up the differences MRP II has made in his work life by saying: "If I know what I need to support manufacturing, I have a reasonable chance to get it. Before, I never knew, so we were always fire-fighting. I would never work for another company that didn't have an MRP II system."

CONCLUSIONS

The results achieved in each of the four areas studied did not vary with the size or purchasing power of the company. Therefore, the effectiveness of supplier scheduling does not appear to be dependent upon "clout." We believe firmly that it's a function of the desire and the understanding of the people in the purchasing department.

One additional conclusion drawn from this study is that companies that stay with the traditional way of purchasing (hard-copy purchase orders) do not increase their on-time delivery performance dramatically and see little or no cost reduction.

Overall, the benefits of effective supplier scheduling include:

- Improved credibility with the plant and suppliers because everyone is working to the same set of numbers.

- Reduced inventory without reducing service to the plant. The right items can be brought in as needed.

- The ability to get planning data to the suppliers. The supplier schedule tells the supplier everything the customer knows about future requirements, both short term and out beyond the supplier's quoted lead time.

- Ninety-five percent or better on-time supplier delivery performance.

- Buyers who can now spend time on value analysis, negotiation, and supplier problems, not expediting.

- Reduced floor expediting because manufacturing supervisors have the correct items when they need them.

- Reduced supplier lead times.

- A more motivated, professional purchasing department.

Appendix C:
The ABCD Checklist

INTRODUCTION

PERSPECTIVE

Are we doing the right things? How well are we doing them? It is hard to imagine two more valuable questions for all managers to be frequently asking. The correct answers to them reflect current levels of performance and reveal future opportunities for improvement.

Finding the right answers, however, requires many more questions. To help you analyze and evaluate your operation, we have constructed a checklist representing what the Oliver Wight Group would ask if appraising your effectiveness in planning and controlling your business, and managing your continuous improvement process.

A good checklist does more than tell you where you are today—An effective checklist helps managers focus on what's required to become more competitive. Periodic use generates a consistent means of determining progress. In addition, it surfaces problems early, which enables the correction process to start earlier. Further, by comparing performance against established benchmarks, people are motivated to do their jobs in a better manner.

EVOLUTION

Ollie Wight created our first ABCD Checklist in 1977. It consisted of 20 questions designed to evaluate a Manufacturing Resource Planning system, MRP II. The questions were grouped into three categories: technical issues to determine whether the design was proper, accuracy of data to determine how reliable the information was; and operational questions to determine how well MRP II was understood and used within the company.

A few years later the original list was expanded to 25 questions. This checklist is the accepted industry standard for measuring MRP II. As a point of reference, we have included this checklist in the *Appendix*. By answering these questions, a manager can objectively grade his company's MRP II system into one of four categories: A, B, C, or D. These levels of potential have served as challenging goals, especially to reach the class A level.

No Longer Just MRP II

Although the checklist has served its purpose admirably, the body of knowledge of MRP II and Distribution Resource Planning has matured and new knowledge of Just-in-Time and Total Quality Control has emerged. More emphasis needs to be placed on the subjects of business planning, manufacturing strategy, sales & operations planning, and the important effort of continuous improvement. The new checklist addresses these changes, although it is not intended to be comprehensive in the area of TQC.

New Checklist

An ideal checklist for executives would be informative, comprehensive, and concise. It is straightforward to do only one, but difficult for one checklist to service all three. As a result, we have created two categories of questions:

• *Overview questions:* They are designed to allow general managers to evaluate whether necessary processes exist and how well they are being used. Whenever possible, they provide industry standards of performance that test the effectiveness of each process.

The 35 questions that make up the overview checklist are separated into four sections. Fifteen questions probe the *planning and control processes*—do they exist and are they being used effectively? The next three questions are aimed at *data management*—how reliable is your planning information? *Continuous improvement processes* are analyzed with the following six questions—are they in place and working correctly? Ten questions are devoted to *performance mea-surements*—are you measuring the right ones and do the levels of performance match the standards established by well-managed com-panies?

There is one other question which does not fit into any of these categories and is the first to be answered—how serious do you and your top management team consider these four issues? Any company lacking a strong commitment to operate the business in a controlled manner and to constantly strive to improve performance will find little value in responding to the other 34 overview questions.

- *Audit questions:* they are designed to provide operating managers with a means for assessing the significant characteristics of each process, checking the vital "how to's" of each process, and analyzing in greater detail how well the processes are being used. For most of the overview questions, several audit questions are listed that will help determine the appropriate answer for each overview question.

Several leading-edge users of MRP II, JIT, and DRP participated in developing and critiquing our new ABCD checklist. More than one expressed concern regarding the volume of audit questions. "My initial reaction when I saw the audit list was, 'wow this is too much.' However, once I started into it, I found that it was very comprehensive and right on target." The number of audit questions was limited only by one criterion—does it contribute to the objective of the review?

HOW TO USE THE CHECKLIST

The two groups of questions, overview and audit, are designed to support each other. The audit questions provide a basis for a "yes/no" response to the overview question. In turn, a manager who wishes to evaluate any overview question with a "no" answer, should turn to the audit questions for more detailed understanding in order to initiate corrective action.

One executive who provided valuable feedback described how he used the combination of overview questions and audit questions: "If I could say 'yes' to most of the audit questions and if we are getting the desired results in the area that overview question is aimed at, then I said, 'yes' to the overview question." This is the correct intent for utilizing the two categories of questions. They boil down to a simple issue: are you getting the desired results? If the results are not evident, it doesn't really count whether you have the right techniques in place, each designed properly, information available when needed, and users working hard. Together, this simply adds up to missed opportunities.

PREREQUISITES

There are a few prerequisites for the new ABCD Checklist to be useful. The first is that the user must be knowledgeable. This means being familiar with the terms and techniques which are referenced and having an adequate understanding as to why the processes are important for the company to operate at a very high standard. Secondly, it assumes that the answers will come from people of "good intentions." Knowledgeable people who are sincerely attempting to be objective can avoid seeing the world through "rosecolored glasses" or being overly critical to the extent that any minor imperfection leads to a negative response.

No matter how well questions may be phrased, we recognize that degrees of interpretation are required to answer them. We also realize that a significant element of judgment is needed before you can say "yes" or "no" to many questions. The combination of interpretation and judgment will, hopefully, lead to healthy internal discussion. The checklist will be productive if companies use it to review why these processes are important, what each process consists of, how the process can contribute to improvements, and how to accomplish these improvements.

MEASURING EFFECTIVENESS

Do you have the right processes and tools installed? How well are you using them? The answers to the overview list provide the basis for determining your effectiveness in the areas covered by the checklist. You can determine whether your company is operating at a Class A, B, C, or D level.

Although there is a quantitative method for judging the appropriate level, there are key qualitative characteristics which distinguish companies:

LEVEL	PLANNING & CONTROL PROCESSES	CONTINUOUS IMPROVEMENT PROCESSES
Class A	Effectively used company-wide; generating significant improvements in customer service, productivity, inventory, and costs.	Continuous improvement has become way-of-life for employees, suppliers, and customers; improved quality, reduced costs, and increased velocity are contributing to a competitive advantage.
Class B	Supported by top management; used by middle management to achieve measurable company improvements.	Most departments participating and active involvement with some suppliers and customers; making substantial contributions in many areas.
Class C	Operated primarily as better methods for ordering materials; contributing to better inventory management.	Processes utilized in limited areas; some departmental improvements.
Class D	Information inaccurate and poorly understood by users; providing little help in running the business.	Processes not established.

To determine the current level for your company, add up the number of "no" answers on the overview list:

- Total of 0 to 3 "no" responses means that you are at a Class A level.
- Total of 4 to 7 "no" answers qualifies for a B level.
- Total of 8 to 10 "no's" means Class C.
- Total of 11 to 14 "no" answers indicates a Class D level.

Our group considers certain performance measurements mandatory to qualify for Class A. Unless there are "yes" answers to questions 31 through 33, reflecting that measurable results have been achieved, we would not judge a company to be at an "A" level. The acid test of applying the right processes and correct techniques is to produce measurable paybacks.

Moreover, before any company can be rated with confidence, there should be at least three months of sustained performance. As we all know, there are periods where everything appears to be working well, but it does not necessarily mean that the company has implemented the right set of tools nor has learned to manage effectively using them. A single point in time is not sufficient to arrive at a firm conclusion.

Even a Class A rating, however, should not be interpreted as achieving full utilization of your company's capabilities. Because it's such a difficult challenge to achieve, any group should take pride in attaining a Class A level and then use this accomplishment as inspiration for further improvements. In fact, all of the Class A companies that we are familiar with tend to be very self-critical. They continue to see what remains, what can be done, and are aggressively pushing forward. "I know we can get a lot better and must." "The emphasis in our business is to give the customer what he wants, when he wants it, with continually lower lead times, and excellent perceived quality. The competition is tough, but so are we." These quotes come from companies that are remarkably good today, but will be even better tomorrow.

OBJECTIVE

Our objective is to help companies become the best they can be. If you find this new ABCD Checklist helps you to ask the right questions and to find the right answers to accomplish this, we have produced a Class A product.

Walter E. Goddard
President
The Oliver Wight Companies

A great number of people made contributions to this checklist, but four individuals deserve special mention: Dick Ling, who was the chairman of this task force, and also George Palmatier, John Sari, and Steve Souza. Without the team effort by this group of people, this checklist never would have been possible.

OVERVIEW QUESTIONS

Asking the right questions to determine your effectiveness in:

• Planning and controlling your business.
• Managing your continual improvement process.

	YES	NO

1. *Management* is committed to use planning and control processes and continuous improvement processes and considers their effective use essential for the survival, growth and general well-being of the company.

☐ ☐

PLANNING AND CONTROL PROCESSES

2. A *strategic planning* process is used to establish the organizational purpose, provide direction for future generations of products and services and develop a competitive strategy which includes a manufacturing strategy statement. A vision statement is used to provide direction to this process. (I.A.)*   ☐ ☐

3. A *business planning* process, driven by the strategic plan, is used to develop and communicate annual financial plans which incorporate input from all operating departments of the company. (I.B.)   ☐ ☐

4. *Sales & operations planning* is the management process that maintains the current operating plan in support of the business plan. The process consists of a formal meeting each month run by the general manager and covers a planning horizon adequate to plan resources effectively. (I.C.)   ☐ ☐

5. A *single set of numbers* is used by all functions with the operating system providing the source data used for *financial planning, reporting, and measurement.* (I.D.)   ☐ ☐

6. *"What if" simulations* are used to evaluate alternative operating plans and to develop contingency plans. (I.E.)   ☐ ☐

YES   NO

7. There is a process for *forecasting* all anticipated demands with sufficient detail and adequate planning horizon to support business planning, sales & operations planning and master production scheduling. Forecast accuracy is measured in order to continuously improve the process. (I.F.)

8. There is a formal *sales planning* process in place with the sales force responsible and accountable for developing and executing the resulting sales plan. Differences between the sales plan and the forecast are reconciled. (I.F.)

9. *Customer order entry and promising* are integrated with the master production scheduling system and inventory data. There are mechanisms for matching incoming orders to forecasts and for handling abnormal demands. (I.F.)

10. The *master production scheduling* process is perpetually managed in order to insure a balance of stability and responsiveness. The master production schedule is reconciled with the production plan resulting from the sales & operations planning process. (I.G.)

11. A *supplier planning and scheduling* process provides visibility for key items covering an adequate planning horizon. (I.H.)

12. There is a *material planning* process which maintains valid schedules and a *material control* process which communicates priorities through a manufacturing schedule, dispatch list, supplier schedule and/or a kanban mechanism. (I.I.)

13. There is a *capacity planning* process using rough-cut capacity planning and, where applicable, capacity requirements planning in which planned capacity, based on demonstrated output, is balanced with required capacity. A *capacity control* process is used to measure and manage factory throughput and queues. (I.J.)

YES   NO

14. All phases of *new product development* are integrated ☐ ☐
with the planning and control system. (I.K.)

15. Where applicable, *engineering activities in support of a* ☐ ☐
customer order are integrated with the planning and
control system. (I.K.)

16. Where applicable, *Distribution Resource Planning* is ☐ ☐
utilized to manage the logistics of distribution. DRP
information is used for sales & operations planning,
master production scheduling, supplier scheduling,
transportation planning and the scheduling of shipping. (I.L.)

DATA MANAGEMENT

17. The planning and control process is supported by a ☐ ☐
properly structured, integrated set of bill of material,
routing and related data. (II.)

18. *Data integrity* is measured against pre-established tol- ☐ ☐
erances and meets accuracy requirements including
the following fundamental standards:
   a. Bills of material, formulas, etc.   98–100%
   b. Routings   95–100%
   c. Inventory records   95–100% (II)

19. There is an effective process for *evaluating, planning* ☐ ☐
and controlling changes to existing products. (II.)

CONTINUOUS IMPROVEMENT PROCESSES

20. The company is committed to a program designed to ☐ ☐
provide appropriate *education* to all employees,
enabling the effective *management of change.* (III.A.)

21. There is an active *employee involvement* program ☐ ☐
designed to improve company operations using the
knowledge and experience of all employees. The program includes cross-training to improve company flexibility and employee security. (III.A.)

	YES	NO

22. There is a company-wide commitment to *continuous improvement* using a "one less at a time" process to stimulate the elimination of non-value adding activities by surfacing, prioritizing and resolving problems. (III.B.) □ □

23. There is a company-wide *total quality improvement process* to insure that the output of each functional area meets or exceeds their customer requirements, internal or external, and which seeks to minimize process and product variation. (III.B.) □ □

24. There is a defined *product development* strategy that, in addition to customer, marketing and technical requirements, considers the issues of manufacturability and involves suppliers when developing new product designs. (III.B.) □ □

25. Strong *"partnership" relationships* with customers and suppliers are being established and the number of suppliers is being reduced. □ □

PERFORMANCE MEASUREMENTS

Planning and Control Process Measurements

26. *Production plan performance* is ± 2% of the monthly plan. □ □

27. *Master production schedule* performance is 95–100% of plan. □ □

28. *Manufacturing schedule performance* is 95–100% of plan. □ □

29. *Engineering schedule performance* is 95–100% of plan. □ □

30. *Supplier delivery performance* is 95–100% of plan. □ □

Company Performance Measurements

	YES	NO

31. *Customer service* □ □
    a. Delivery to first promise and/or line item fill rate is 95–100%
    b. An objective for delivery to customer request date exists and performance against the objective is measured.

32. *Quality* □ □
    Objectives are established, performance is measured and goals are achieved for
    a. Production defects.
    b. Supplier defects.
    c. Cost of quality.

33. *Cost* □ □
    Performance against set objectives is measured for reducing total costs including labor, overhead, material, distribution and transportation costs where applicable.

34. *Velocity* for all departments is measured and is improving for: □ □
    a. Delivery times from suppliers.
    b. Manufacturing lead times.
    c. Delivery times to customers.
    d. Design time.

35. *Management* uses performance measurements, including this checklist, to continuously stimulate improvements, not simply as a rating device. □ □

*Refer to audit list for corresponding questions.

OVERVIEW QUESTIONS SUMMARY SHEET

	Yes	No
1. Management Commitment	____	____

Planning and Control Processes

	Yes	No
2. Strategic Planning	____	____
3. Business Planning	____	____
4. Sales & Operations Planning	____	____
5. Single set of numbers	____	____
6. "What if" Simulations	____	____
7. Forecasts that are measured	____	____
8. Sales Plans	____	____
9. Integrated Customer Order Entry and Promising	____	____
10. Master Production Scheduling	____	____
11. Supplier Planning and Control	____	____
12. Material Planning and Control	____	____
13. Capacity Planning and Control	____	____
14. New Product Development	____	____
15. Engineering integrated	____	____
16. Distribution Resource Planning	____	____

Data Management

	Yes	No
17. Integrated BOM and Routing	____	____
18. Data Accuracy	____	____
BOM 98–100%		
Routing 95–100%		
Inventory records 95–100%		
19. Product change control	____	____

Continuous Improvement

	Yes	No
20. Employee education	____	____
21. Employee involvement	____	____
22. One less at a time	____	____
23. Total Quality Improvement process	____	____
24. Product development strategy	____	____
25. Partner relationship with customers	____	____

Performance Measurements

Planning and Control Process Measurements

	Yes	No
26. Production Planning performance ± 2%	____	____
27. Master Production Schedule performance 95–100%	____	____
28. Manufacturing schedule performance 95–100%	____	____
29. Engineering schedule performance 95–100%	____	____
30. Supplier delivery performance 95–100%	____	____

Company Performance Measurements

	Yes	No
31. Customer service delivery to promise 95–100%	____	____
32. Quality performance measured	____	____
33. Cost performance measured	____	____
34. Velocity performance measured	____	____
35. Management uses measurements for improvements	____	____

AUDIT QUESTIONS OUTLINE

I. PLANNING AND CONTROL PROCESSES

A. Strategic Planning
B. Business Planning
C. Sales & Operations Planning
D. Financial Planning, Reporting and Measurements
E. "What if?" Simulations
F. Demand Management—Forecasting, Sales Planning and Order Entry and Promising
G. Master Production Scheduling
H. Supplier/Purchasing Management
I.   Material Planning and Material Control
J.   Capacity Planning and Capacity Control
K. New Product/Engineered Product Planning and Control.
L. Distribution Resource Planning

II. DATA MANAGEMENT

III. CONTINUOUS IMPROVEMENT PROCESSES

A. Continuous Improvement
B. Education and Managing Change

IV. PERFORMANCE MEASUREMENTS

AUDIT QUESTIONS

I. PLANNING AND CONTROL PROCESSES

A. *Strategic Planning*

	YES	NO
1. Strategy development includes and integrates all functional areas of the company; finance, marketing, manufacturing, product development and all support organizations.	☐	☐
2. The strategic plan includes a statement of purpose or mission.	☐	☐
3. The strategic planning process addresses both internal factors (conditions you can control) and external factors (conditions you cannot control) that significantly affect the business.	☐	☐
4. The competitive strategy is based on a clear understanding of the existing and anticipated competition.	☐	☐
5. The strategic plan's horizon covers the time required, typically 3–5 years, to implement major changes in strategy and major changes to the market.	☐	☐
6. The strategic planning process occurs at least annually with review and update at least every three months.	☐	☐
7. The competitive strategy is clearly defined, to provide adequate direction for all functional areas of the company. Each functional area uses this to develop its own supporting strategy statement.	☐	☐
8. The strategic plan is communicated to the appropriate people to be used as a guide in developing the business plan and sales & operations plan.	☐	☐

B. Business Planning

	YES	NO

1. The business plan provides specific direction regarding market share, financial performance, marketing/manufacturing strategy, new product development, customer service levels and desired inventory levels to be used in the sales & operations planning process. ☐ ☐
2. The business plan horizon is long enough to develop the company's sales & operations planning process. ☐ ☐
3. The business plan is developed at least annually and reviewed at least quarterly or more frequently if business conditions change significantly. ☐ ☐
4. The underlying assumptions used to develop the strategic and business plans are adequately documented. ☐ ☐
5. A mechanism is in place to check that the sales & operations plan is synchronized with the business plan. ☐ ☐
6. The business plan provides a detailed financial plan by department. Department managers participate in development of the business plan. Department managers review progress against the business plan on a monthly basis. ☐ ☐

C. Sales & Operations Planning

1. There is a concise written sales & operations planning policy that covers the purpose, process and participants in the process. ☐ ☐
2. Sales & operations planning is truly a process and not just a meeting. There is a sequence of steps that are laid out and followed. ☐ ☐
3. The meeting dates are set well ahead to avoid schedule conflicts. In case of an emergency, the department manager is represented by someone who is empowered to speak for the department. ☐ ☐
4. A formal agenda is circulated prior to the meeting. ☐ ☐

	YES	NO

5. The plans are reviewed by product family units of measure that communicate most effectively. ☐ ☐

6. The new product development schedule is reviewed at the sales & operations planning meeting. ☐ ☐

7. All participants attend the sales & operations planning meeting prepared. There are preliminary meetings by each department in preparation for the sales & operations planning meeting. ☐ ☐

8. The presentation of information includes a review of both past performances and future plans for: sales, production, inventory, backlog, shipments and new product activity. ☐ ☐

9. Inventory and/or delivery leadtime (backlog) strategies are reviewed each month as part of the process. ☐ ☐

10. There is a process of reviewing and documenting assumptions about business and the marketplace. This is to enhance the understanding of the business and represents the basis for future projections. ☐ ☐

11. Sales & operations planning is an action process. Conflicts are resolved and decisions are made and communicated. ☐ ☐

12. Any changes—large and/or unanticipated, are communicated to other departments prior to the meeting. ☐ ☐

13. Minutes of the meeting are circulated immediately after the meeting. ☐ ☐

14. The mechanism is in place to insure that aggregate sales plans agree with detailed sales plans by item and by market segment or territory. There is a consensus from sales, marketing and operating management. ☐ ☐

15. Time fences have been established as guidelines for managing changes. In the near-term, there is an effort to minimize the changes in order to gain the benefits of stability. In the mid-term range, changes up or down are expected but are reviewed to insure they can be executed. In long-term, less precision is expected but direction is established. ☐ ☐

	YES	NO
16. Tolerances are established to determine acceptable performance for: sales, engineering, finance and production. They are reviewed and updated. Accountability is clearly established.	☐	☐
17. The production plan is the driver of the master schedule and is supported by a procedure which defines a summarization to insure that they are in agreement.	☐	☐
18. There is an ongoing critique of the process.	☐	☐

D. Financial Planning, Reporting and Measurements

	YES	NO
1. The financial projections developed in the sales & operations planning process are linked to the company's financial plans. When financial projections differ from the financial plans contained in the business plan, the differences are reconciled and either the sales & operations plan or the business plan updated.	☐	☐
2. The finance department uses the same data on sales, shipments, etc. as other departments to measure performance.	☐	☐
3. The finance department recognizes limits of traditional performance measurements, particularly those related to overhead allocation, and understands when and how those methods may produce misleading or incorrect data.	☐	☐
4. All financial systems (billing, accounts payable, cost accounting, purchasing, receiving, etc.) are fully integrated with all transaction systems.	☐	☐
5. Accounts payable, purchasing, and receiving tie to material receipt transactions.	☐	☐
6. Material receipt transactions are costed and recorded in a general ledger for inventory valuation purposes.	☐	☐
7. Where applicable, labor reporting and material issues are used to determine the cost of the product and to update the daily dispatch lists.	☐	☐

YES NO

8. Work order closing transactions, if applicable, are used to generate movement of inventory from one account to another in the general ledger and also to trigger variance reports for cost accounting purposes. ☐ ☐

9. Customer orders shipment transactions drive the updating of finished goods inventory and the billing system at the same time. ☐ ☐

10. Financial measurements, particularly those related to overhead allocation, which impede improvements that would result from sound Just-in-Time practices have been identified and are being or have been eliminated. ☐ ☐

11. Simulation tools are actively used to convert operating data into financial data quickly for the purpose of simulation testing, decision making and contingency planning. A chart is available that shows how key MRP II reports translate into key financial reports. ☐ ☐

12. Finance department is proactive in simplifying the cost accounting system and eliminating non-value adding activities. ☐ ☐

E. *"What If?" Simulations*

1. There is a computer-based process supporting sales & operations planning which permits the evaluation of various levels of demand, supply, production, inventory and/or backlogs. ☐ ☐

2. There is a "what if?" process supporting customer order entry and promising used to determine the effects of making customer promises. ☐ ☐

3. Rough-cut capacity planning is used to evaluate the impact on critical resources of alternative production and master production schedule plans. "Off-load" strategies and suppliers are in place. ☐ ☐

4. Where applicable, capacity requirements planning is used to evaluate detailed capacity constraints when planning and budgeting labor and equipment needs. ☐ ☐

YES   NO

5. Material Requirements Planning and Distribution Requirements Planning, where applicable, are used periodically to evaluate lot size alternatives and evaluate the resultant impact on inventory levels.   ☐ ☐

F. Demand Management—Forecasting, Sales Planning and Order Entry Promising

1. There is clear accountability for developing the forecast, and the importance of this effort is reflected in the organization and reporting relationship of the forecasting function.   ☐ ☐
2. The forecaster understands the product, the customer base, the marketplace and the manufacturing system.   ☐ ☐
3. The sales force understands the impact of sales planning on the company's ability to satisfy its customers.   ☐ ☐
4. All demands are included in the forecast, e.g. spares, samples, internal use, interplant demand, etc.   ☐ ☐
5. Available statistical forecasting tools are utilized where applicable.   ☐ ☐
6. Actual sales are measured against sales plans. Measurements are broken down into sales responsibility areas.   ☐ ☐
7. The sales planning process is designed in such a way as to minimize the administrative impact for the sales force.   ☐ ☐
8. The incentives of the sales compensation system are effective and do not inject bias into the sales planning and forecasting.   ☐ ☐
9. The sales force is actively pursuing customer-linking with their customers. They are working with their customer's planning systems and communicating this information to the company.   ☐ ☐
10. Spare parts and other lower-level demands are handled with a forecasting system and appropriate order entry   ☐ ☐

YES   NO

mechanism that introduces the demands at the right level in the MRP process.

11. Abnormal demand (both active and history) is coded properly in the data base. ☐ ☐

12. The order promising function has access to appropriate information, such as available-to-promise (ATP), to insure that good promises can be made. ☐ ☐

13. Sales and marketing participate in developing appropriate time fences for managing change. ☐ ☐

14. Aggregate forecasts are reconciled with the sales plan. ☐ ☐

15. Detail forecasts are reconciled with aggregate forecasts and communicated to the master production scheduler and sales force. ☐ ☐

16. The significant assumptions underlying the forecast are documented. They are reviewed at least monthly and updated as market conditions change. ☐ ☐

17. The forecaster participates in the product management and product development processes, including product structuring. ☐ ☐

18. Sales participates with marketing, forecasting and manufacturing in a demand planning meeting to prepare for each sales & operations planning meeting. A system is in use to communicate customer intelligence information to forecasting. ☐ ☐

19. Sales areas are provided with useful feedback regarding their performance to plan at least monthly. Sales plans are stated so that they are meaningful to the sales force, yet translate into the sales & operations process. ☐ ☐

20. The assumptions underlying the sales plan are documented. They are reviewed on a regular basis and changed as necessary. ☐ ☐

21. There is a process in place for identifying abnormal demands and dealing with them. ☐ ☐

22. Aggregate and detailed measurements are used, and goals are established and measured with continuous improvement as the primary objective. ☐ ☐

	YES	NO

23. Customer orders are processed on a timely basis. The □ □
backlog of customer orders is measured and managed.
24. Order entry errors are measured and managed to an □ □
acceptable level.
25. The number of customer initiated sales change orders □ □
is measured and managed to an acceptable level.

G. Master Production Scheduling

1. Accountability for maintaining the master schedule is □ □
clear. The importance of master scheduling is reflected
in the organization and reporting relationship of the
master scheduling function.
2. Master schedulers understand the product, manufac- □ □
turing process and the manufacturing planning and
control system.
3. The master scheduler participates in and provides □ □
important situation information to the sales & operations planning process.
4. The master scheduler reacts to feedback which identi- □ □
fies material and capacity availability problems and
initiates the problem resolution process.
5. Planning bills of material, if needed, are maintained □ □
jointly by the master scheduler and sales and marketing.
6. A written master schedule policy is followed to moni- □ □
tor stability and responsiveness, goals are established
and measured.
7. Policy governs the use of safety stock/option overplan- □ □
ning used to increase responsiveness and compensate
for inconsistent supply or variable demand.
8. The master schedule is summarized appropriately and □ □
reconciled with the agreed-to rate of manufacture (production plan) from the sales & operations planning
process.
9. All levels of master scheduled items are identified and □ □
master scheduled.

	YES	NO

10. The master schedule is in weekly, daily or smaller periods and is replanned at least weekly. ☐ ☐

11. The master schedule is "firmed up" over a sufficient horizon using time fences and firm planned orders. Guidelines for this horizon include a) cumulative lead time, b) current backlog, c) the need for stability of operations. ☐ ☐

12. The bill of material structure supports the forecasting/ master scheduling process. ☐ ☐

13. Forecast consumption processes are used to prevent planning nervousness. ☐ ☐

14. The alternative approaches used with planning bills of material to develop production forecasts for master scheduled items are well understood and an appropriate process is used. ☐ ☐

15. Available to promise product/capacity information is used for customer promises. ☐ ☐

16. Rough-cut capacity planning, or its equivalent, is used to evaluate the impact on critical resources of significant master schedule changes. Demonstrated capacity is measured and compared to required capacity. ☐ ☐

17. A finishing/final assembly mechanism coordinated with the master schedule is used to replenish finished goods or schedule customer orders to completion. ☐ ☐

18. Mixed-model master scheduling is being pursued where applicable. ☐ ☐

19. Master schedule changes within the "firmed up" period (closest time fence) are measured and reviewed for cause. ☐ ☐

20. A weekly master schedule communications meeting exists and is attended by all using functions. ☐ ☐

H. Supplier Planning and Control

1. At least 80% of the suppliers have been educated in MRP II and/or JIT and understand the vendor scheduling process. ☐ ☐

	YES	NO

2. Suppliers understand the principles of vendor scheduling and agree to plan raw material and capacity requirements to meet the total requirements displayed on the vendor schedule. ☐ ☐

3. There is a clear policy statement as to the responsibilities of the vendor scheduler and buyer, including at what point each becomes involved in problem resolution. ☐ ☐

4. The vendor schedule displays scheduled receipts and planned orders over the planning horizon for all parts a supplier provides. ☐ ☐

5. Time periods on the vendor schedule are weeks or smaller for at least the first four weeks displayed. ☐ ☐

6. The vendor schedulers and/or buyers meet with planners as frequently as required to maintain a valid schedule. ☐ ☐

7. Scheduled receipts cover the supplier's quoted lead-time and planned orders out beyond the quoted lead-times give the vendor future visibility. ☐ ☐

8. Commitment zones are established in the vendor schedule representing firm commitments, material commitments and capacity planning commitments. ☐ ☐

9. The suppliers understand the principle behind "silence is approval" and agree to notify the buyer in advance if a due date will be missed. ☐ ☐

10. Vendor schedules are communicated to vendors at least weekly. ☐ ☐

11. For non-vendor schedule items, 95% of purchase orders are released with full lead time. ☐ ☐

12. There is a set of criteria which defines "key" items which are planned and scheduled through a vendor scheduling process. Considerations include factors such as 80% of purchase content, long lead times, critical items, etc. Some items such as miscellaneous hardware might be excluded. ☐ ☐

YES   NO

13. Simultaneous improvement has been achieved in at   ☐   ☐
    least three of the following areas: inventory turnover,
    customer service, freight costs, obsolescence and
    inter-distribution center transfers.

I. Material Planning and Material Control

1. Material planners understand the product, the manu-   ☐   ☐
   facturing process, and the manufacturing planning and
   control system.
2. Everyone, including planners, production supervisors,   ☐   ☐
   buyers, etc., operates under the "silence is approval"
   principle and is responsible to feed back schedule
   problems which cannot be resolved.
3. Planners are responsible for maintaining, periodically   ☐   ☐
   reviewing and analyzing the accuracy and validity of
   all appropriate planning parameters, which may
   include: order quantities or lot sizes; lead times,
   queues; safety stocks; etc.
4. Production supervisors and buyers understand and use   ☐   ☐
   the system and are accountable for maintaining data
   validity including point-of-use inventory, assigning
   planning parameters and schedule or order data.
5. There are formal review meetings between planning,   ☐   ☐
   production and purchasing.
6. The informal shortage list has been eliminated and is   ☐   ☐
   no longer the priority setting mechanism.

Material Requirements Planning, if applicable

7. MRP time periods are weekly or smaller to provide   ☐   ☐
   appropriate resolution of priorities.
8. The MRP system is run as frequently as required to   ☐   ☐
   maintain valid schedules. Daily or "on-line" may be
   required. Weekly processing is normally minimum.

YES   NO

9. The system uses standard logic to generate action/ ☐ ☐
exception messages including Need to Release Order,
Need to Reschedule Order, Need to Cancel Order, Due
Date or Release Past Due.

10. The system has a firm planned order capability which ☐ ☐
is used, when necessary, to override the suggested
plan.

11. The system has, at a minimum, single-level pegging ☐ ☐
capability and planners use it to identify the source of
demand when resolving scheduling problems.

12. The system has an effective component availabil- ☐ ☐
ity checking mechanism and the planners use it to
determine the feasibility of releasing an order or
schedule.

13. The system includes the capability to alter the bill of ☐ ☐
material for an individual order in order to handle
temporary substitutions, etc. where applicable.

14. A bottom-up re-planning process is used to reconcile ☐ ☐
problems and exception messages.

15. All action/exception messages are prioritized and iden- ☐ ☐
tified problems are acted upon when appropriate
before the next MRP run.

16. The number of action/exception messages for each ☐ ☐
planner is monitored for activity and trends.

17. The volume of reschedules is tracked to determine the ☐ ☐
stability of the plan and the causes of excessive
rescheduling activity.

18. Orders are released with full material availability and ☐ ☐
full lead time 95–100% of the time.

Shop Floor Control, if applicable

19. The system includes the back-scheduling capability to ☐ ☐
create start and due dates on a work order and operations within a routing.

	YES	NO
20. The system includes the capability to modify all start and due dates on a work order and operations within a routing.	☐	☐
21. The system includes the capability to report status by operation.	☐	☐
22. An anticipated delay reporting process is used to maintain due date validity.	☐	☐
23. The system includes a dispatch list which shows by work center: part number, order number, operation number, operation start and complete date and order due date.	☐	☐
24. Production management is accountable to meet operation due dates.	☐	☐
25. The dispatch list is the only priority tool and operation due and start dates are the only priority techniques used.	☐	☐

Kanban processes, if applicable

1. Open kanbans or kanban signals authorize work to be performed.	☐	☐
2. Operators have the authority to stop production for quality reasons.	☐	☐

J. Capacity Planning and Capacity Control

1. Capacity planning is well understood by all appropriate personnel and used to plan labor and machinery requirements.	☐	☐
2. The production supervisor is held accountable for monitoring capacity planning parameters (e.g. planned capacity, number of workers and/or machines, number of shifts).	☐	☐

	YES	NO
3. Production and capacity planners meet at least weekly to resolve capacity issues.	☐	☐
4. All sources of demand are considered in developing the capacity requirement.	☐	☐
5. When appropriate, engineering and vendor capacity information is included in the capacity management process.	☐	☐
6. A "Load Factor" is maintained and used in projecting capacity.	☐	☐
7. Corrective action is taken to address overdue capacity requirements.	☐	☐
8. Process includes appropriate productivity analysis.	☐	☐

Capacity Requirements Planning, where applicable

	YES	NO
9. System produces capacity requirements summary report by work center and a detailed capacity report.	☐	☐
10. Capacity planning parameters are 95–100% accurate.	☐	☐
11. Process includes variance analysis of planned and actual input, output and queue levels (Input/Output Report).	☐	☐

K. New Product/Engineered Product Planning and Control

1. There is an engineering department policy of design review with manufacturing, marketing and quality assurance during the crucial stages of design for each new development project.	☐	☐
2. There is an engineering department policy on commonality of components in new designs with those in existing designs when practical.	☐	☐
3. There is a detailed product development schedule with authorization and funding to support each new product development project.	☐	☐

YES   NO

4. Planning and control process methodology such as rough-cut capacity and capacity requirements planning are used to evaluate new product resource requirements. ☐ ☐

5. The new product development schedule is reviewed at the sales & operations planning meeting and the formal system is used to communicate changes. ☐ ☐

6. Once a new product has been authorized, it is included in the sales & operations planning and master scheduling process. ☐ ☐

7. Advanced manufacturing team disciplines are utilized to plan, accelerate the transition and improve the integration of new product development. ☐ ☐

8. Managers within the engineering department understand the planning and scheduling system. They have accepted the responsibility for effectively using it. ☐ ☐

9. For selected engineering work centers, predictions of required capacity and measurements of actual output are provided by the planning and scheduling system. Managers use this information to insure that a proper balance is achieved between the two. ☐ ☐

10. If required, rough-cut capacity planning representing engineering's capabilities is used to provide information for the sales & operations planning and master scheduling processes. ☐ ☐

11. Priority planning for engineering work is linked with the master production schedule. Up-to-date priorities are communicated to the work centers within the engineering department. ☐ ☐

12. On-time deliveries from engineering are measured. When schedules cannot be met, feedback is initiated so consequences can be determined. Causes of these delays are tracked and revised completion dates are provided. ☐ ☐

13. Sufficient capacity in engineering is allocated to encourage and implement product improvements sug- ☐ ☐

gested by customers, sales, marketing, manufacturing, quality assurance, suppliers and others.

L. Distribution Resource Planning

1. There is a concise written Distribution Resource Plan- ☐   ☐
   ning policy which covers purpose, process and participants.
2. Distribution requirements are considered and recon- ☐   ☐
   ciled through the sales & operations planning and master scheduling processes.
3. The distribution network includes all products at each ☐   ☐
   distribution center.
4. Forecasts are available for each stockkeeping unit in ☐   ☐
   each distribution center.
5. Time periods for DRP are weeks or smaller. ☐   ☐
6. Distribution Resource Planning is run weekly or more ☐   ☐
   frequently.
7. The DRP system includes the following capabilities: ☐   ☐
   a. Firm planned orders
   b. Pegging capabilities
   c. Customer orders promised for future deliveries in
      addition to forecasts.
   d. Includes backorders in the netting logic
   e. The ability to maintain and change inventory rec-
      ords, location records and scheduled receipts.
   f. Vendor scheduling in order to provide adequate
      visibility to outside suppliers
   g. Rescheduling messages
8. The system provides pertinent information for trans- ☐   ☐
   portation planning in order to be responsive to the
   needs of the branches as well as to reduce transportation costs.
9. The system includes a shipping schedule which reflects ☐   ☐
   the continuing needs to keep costs at a minimum while

YES   NO

at the same time satisfies established loading and shipping capacity.

10. There are regular (at least monthly) forecast planning   ☐   ☐
meetings including at least the distribution, marketing, purchasing and planning functions.

11. The inventory records at each distribution center are   ☐   ☐
95% accurate or better.

12. The education process initially covered 80% of distri-   ☐   ☐
bution employees and includes an on-going education
program aimed at continually improving the use of the
system.

13. Simultaneous improvement has been achieved in at   ☐   ☐
least three of the following areas: inventory turnover,
customer service, freight costs, obsolescence and
inter-distribution center transfers.

II. DATA MANAGEMENT

1. Cycle counting procedures are used to identify and   ☐   ☐
resolve inventory errors and measure inventory accuracy.

2. The cycle counting process has replaced the periodic   ☐   ☐
physical inventory.

3. Regular audits for accuracy are made for: ☐   ☐
   a. Bills of material
   b. Routing and work center data
   c. Item master planning data

4. Responsibility and accountability for developing and   ☐   ☐
maintaining bills of material are clearly defined in
written policy.

5. All bill of material user functions participate in struc-   ☐   ☐
turing the bill of material.

6. Bills of material are properly structured, represent the   ☐   ☐
way products are built, and support the planning and
control processes.

	YES	NO

7. Bills of material are periodically reviewed in an effort to make them shallower and flatter enabling shorter manufacturing lead times. ☐ ☐

8. An engineering change review board meets regularly to review major engineering change requests. ☐ ☐

9. Finance uses the bill of material to cost the product. ☐ ☐

10. There is an ongoing effort to reduce the number of part numbers. (component standardization) ☐ ☐

11. Manufactured and purchased scheduled receipt data is regularly audited by user functions for accuracy of quantity of an order, timing and order record details. ☐ ☐

12. Work centers are appropriately defined to enable control of priorities and capacities. ☐ ☐

13. A written engineering change policy is followed to insure consistent and timely engineering changes. ☐ ☐

14. The importance of engineering change coordination is reflected in the organization and reporting relationship of the engineering change coordination function. ☐ ☐

15. Appropriate measurements are used to insure engineering changes are carried out in a timely manner. ☐ ☐

16. Rough cut capacity planning is used to evaluate and schedule significant engineering changes. ☐ ☐

17. Appropriate planning and control techniques are used to manage phase-in and phase-out of materials in a cost effective manner. ☐ ☐

18. All necessary items such as tools, fixtures, sales materials, inspection procedures, process instructions, etc are properly managed through engineering change procedures. ☐ ☐

III. CONTINUOUS IMPROVEMENT PROCESS

A. *Education and Training*

Principles

	YES	NO
1. Management attitude and actions demonstrate a commitment to fully educate and train people prior to project implementation.	☐	☐
2. The education program is based on the principles of behavior change in an organization, rather than just transferring a series of facts about how business tools like MRP II and Just-in-Time work.	☐	☐
3. Education is a participative process, rather than a one directional flow—from the top of the organization chart to the bottom. The educational process recognizes people at all levels as experts in their particular area, communicates objectives, and fully involves people in the process of changing their jobs.	☐	☐
4. An ongoing program of education and training is used to refine and improve the use of business tools like MRP II and Just-in-Time. This starts with an audit or assessment of strengths and weaknesses which is used to develop a tailored program of education and training focusing on the areas that will provide the most benefit to the business.	☐	☐

Specific Actions—Initial Education

	YES	NO
5. Active visible top management leadership, peer confirmation, immersion, and credibility. The executive management team (general manager and staff) and the operating management team (department heads reporting to the general manager's staff) have all been through an education session, taught by a credible live	☐	☐

instructor with a track record for success, and including an appropriate level of detail for their respective roles.

6. Ownership. All members of the executive management and operating management teams participate in the cost justification, and agree to be held accountable for the results. The costs and benefits are included into the normal management by objectives process, and the employee appraisal or review process. In other words, there is specific and individual accountability for results.

7. Consensus. As a result of the education process and the cost justification process, the management group has reached a consensus on how they want to run the business, and how it is different from the way the business is currently run.

8. Project management. A project team, executive steering committee, full-time project leader, and detailed implementation plan all exist. The plan identifies the tasks, sequence of tasks, responsibility for tasks, resources required, education required, and due dates for completion of tasks. The project leader and a member of the executive management team are responsible for the creation of this plan. Both have been educated sufficiently to do this work.

9. Line accountability, peer confirmation, and repetition. Two series of business meetings are begun. One for the general manager and staff, the other for the department head team. The objective is to, subject by subject, examine how the business is currently being run in this area (at an appropriate level of detail for the group), determine how the group wants to run this aspect of the business, and list the action items to get there.

At the conclusion of the department head meetings, each manager leads a series of business meetings for his or her department. Again, these meetings go, sub-

YES   NO

ject by subject, over how the business is currently being run, how the group wants to run this aspect of the business, and creates a list of action items. The list of subjects covered, and the depth of coverage will vary by group.

10. Development of in-house experts. In parallel with the business meetings, appropriate people are selected and educated in the technical knowledge needed to support the implementation and operation of the new systems, tools, and techniques.

11. Communicating to suppliers and customers. Appropriate education programs are designed and implemented to allow closer and more effective working relationships with suppliers and customers.

12. Training. Training on how to use the software is done either as part of the series of business meetings mentioned above (preferred), or in a separate stand-alone process.

Specific Actions—Ongoing Education

13. New people. A continuing program of education and training is available for people new to the company, and also for people who have been promoted or transferred to new responsibilities.

14. Improvement and refinement. Each year, an audit or assessment is done to evaluate the specific strengths and weaknesses in using business tools like MRP II and Just-in-Time. This assessment is used to develop a tailored education and training program to focus on the tailored education and training program to focus on the areas that will provide the most benefit to the business. This includes selecting people for some education taught by a credible live instructor, as well as a shortened series of business meetings following the pattern explained above for initial education. This includes the education of suppliers and customers.

YES   NO

15. Cross training. An on-going program of employee □ □
cross training is in place.
16. Participation in relevant professional organizations is □ □
encouraged and supported.

B. Continuous Improvement

Continuous improvement activities are significantly improving performance in the following areas:

Manufacturing Processes

1. Setup and changeover times are being reduced, □ □
enabling manufacturing lot sizes (order quantities) to
be economically reduced.
2. Material handling costs are being reduced. □ □
3. Travel distances that manufactured parts move in the □ □
conversion of raw materials to finished product are
being reduced.
4. Materials are stored primarily at the point-of-use □ □
rather than in stockrooms where applicable.
5. Facilities required to economically receive, produce □ □
and ship product are continuously being made more
cost effective.
6. Plant layouts are being revised to simplify and improve □ □
the physical flow through the factory.
7. Equipment is selected based on how well it will con- □ □
tribute to improved quality, minimum lot sizes, fast
setups and needed flexibility.
8. Functional work centers are being replaced by cells □ □
where appropriate.
9. Simplified manufacturing processes provide visual □ □
controls and problems which surface are quickly identified and resolved.

	YES	NO
10. Good housekeeping is being pursued.	☐	☐
11. Shipment evenness is measured daily. Production is stopped when the shift's quota is met and the remaining time is used to improve the process.	☐	☐
12. An active maintenance program is available for equipment and tooling.	☐	☐
13. Preventative maintenance is improving.	☐	☐

Quality Processes

1. Statistical Process Control is extensively used.	☐	☐
2. Separate inspection activities are being eliminated. Quality is the in-process responsibility of the operator.	☐	☐
3. Quality at the source is being increased.	☐	☐
4. A formal and active program exists to reduce scrap, shrinkage and rework and to increase yield.	☐	☐

Supplier Relationships

1. Just-in-Time/Total Quality Control is being linked throughout the supplier base.	☐	☐
2. Vendors are being qualified to reduce incoming inspection and verification of counts.	☐	☐
3. Order quantities or purchased parts are being economically reduced, resulting in more frequent deliveries from suppliers.	☐	☐
4. Transportation costs from suppliers are being reduced.	☐	☐
5. Materials go from dock to point-of-use rather than from dock-to-stock.	☐	☐
6. Vendor scheduling aimed at providing adequate visibility to suppliers is in place.	☐	☐
7. Suppliers are included on new product design.	☐	☐

People Relationships

YES   NO

1. Improved communications between management and workers is stated as an important company objective and is occurring. ☐ ☐

2. An active training and retraining program for all employees is in place. Its objectives include encouraging the thinking worker, flexibility, employment stability and meeting future needs. ☐ ☐

3. Group meetings are being conducted on a daily basis to work on identifying problems that prevent more economical production of high quality parts. ☐ ☐

4. All functional departments (engineering, purchasing, marketing, manufacturing, etc.) are part of the problem-solving teams. ☐ ☐

5. A process to help front-line supervisors become coaches/facilitators rather than simply bosses is in place. ☐ ☐

6. The personnel department is actively working on the new skills required to support the continuous improvement process and is contributing to the retraining of existing people. ☐ ☐

7. The number of job descriptions in the factory is being reduced. ☐ ☐

8. Operators have the freedom to stop production whenever trouble occurs in order to correct problems and/or improve the process. ☐ ☐

9. Operators take the initiative to move to the point of need. ☐ ☐

10. Compensation is performance based and moving toward a group contribution basis while individual piece-part pay is being minimized. ☐ ☐

Planning and Control Processes

1. A visible process of continuously reducing lead times, order quantities, safety stock, queues and other contributors to inventory is being pursued. A recom- ☐ ☐

YES   NO

mended method for this is reducing the number of "kanbans" which authorize the production and/or purchase or items.

2. Mixed-model scheduling from master scheduling through the entire manufacturing process into purchasing is being pursued. ☐ ☐

3. Daily production rates are being emphasized rather than the conventional "build by batches". ☐ ☐

4. The physical movement of materials within the factory is triggered by a need basis, e.g. a kanban signal. ☐ ☐

5. Bills of material are being flattened. ☐ ☐

6. Routings are being flattened. ☐ ☐

7. Work orders are either being eliminated or simplified. ☐ ☐

8. Paperwork and transactions in the factory and other supporting departments are being reduced or simplified. ☐ ☐

9. Data is available to operational people in a timely manner and a concurrent goal exists to reduce the number of transactions and unnecessary reports. ☐ ☐

Financial Processes

1. Accounting procedures are being simplified, e.g. groupings of direct and indirect operators, labor collection by exception, etc. ☐ ☐

2. Overhead is being allocated on total cost or split between material and "other" rather than simply on direct labor. ☐ ☐

3. The use of traditional labor efficiency and machine utilization to measure performance is being de-emphasized to avoid conflicts with continuous improvement motivation. ☐ ☐

Design Processes

1. Design engineering is part of the company team charged with the responsibility of creating high quality products that are inexpensive to make. ☐ ☐

	YES	NO
2. Design engineering is working closely with manufacturing and purchasing in order to implement the common company objectives.	☐	☐
3. The number of part numbers is being reduced through standardization.	☐	☐
4. An ongoing effort to increase the reliability of on-time completions and decrease lead times within engineering is highly visible, and is generating improvements.	☐	☐
5. Processes to improve "manufacturability" such as "design for assembly" are in use.	☐	☐
6. Engineering changes are analyzed using Total Quality Control methods to reduce their number and associated costs.	☐	☐

Sales, Marketing and Customer Processes

1. Marketing and sales view Just-in-Time/Total Quality Control as a competitive weapon in the marketplace.	☐	☐
2. Marketing and sales are actively working with customers to integrate their needs with the company's manufacturing activities.	☐	☐
3. Just-in-Time/Total Quality Control are being linked throughout the customer base where applicable.	☐	☐
4. Transportation costs to distribution centers and customers are being reduced while responsiveness is increasing.	☐	☐

IV. PERFORMANCE MEASUREMENTS

In addition to the industry standard performance measurements included in the overview questions, listed below are potentially helpful performance measurements to measure progress vs. pre-established targets:

Aggregate Performance Measurements

YES NO

1. Business Plan (return on investment, return on assets, ☐ ☐ total revenue and by product family, total profitability and by product family, market share by product family, by region, etc.).
2. Sales Plan (actual vs. plan, by family, by month, etc.). ☐ ☐
3. Production Plan (actual vs. plan, by family, by month, ☐ ☐ etc.).
4. Inventory (by product line, by finished goods, raw ☐ ☐ material, work-in-process, etc., by actual dollar level and/or turnover rate vs. current sales patterns, etc.).
5. Productivity (output per employee, output per factory ☐ ☐ employee, output per department, work center/line, work cell, level of overtime, level of subcontract, etc.).
6. Customer Service (backorder levels, number of weeks ☐ ☐ aging of overdue customer orders, number of involuntary reschedules of customer orders, by product line, etc.).
7. Quality (level of rework, level of return to vendor, by ☐ ☐ product line, by department, by vendor, by commodity type; level of customer returns, level of field repairs, by product line etc.).

Detail Performance Measurements

Forecasting and Sales Planning

1. Item accuracy (by week or by month, within pre- ☐ ☐ established tolerances, by product family, by geographic region, by responsible Sales/Marketing personnel).
2. Family accuracy (within pre-established tolerances, by ☐ ☐ week or by month, by geographic region, etc.).

Master Production Scheduling

3. Past due (percentage of current output rate, percentage ☐ ☐ of current in-process aging, etc.).

YES   NO

4. Percent schedule changes within near (firm) horizon. ☐ ☐
5. Performance vs. schedule in Finishing/Final Assembly ☐ ☐
per customer order, where appropriate.

Customer Order Entry and Promising

6. Administration customer order promising time (order ☐ ☐
entry, credit check, document generation and delivery,
etc.).
7. Shipment processing time (custom packaging, crating, ☐ ☐
loading, shipping, etc.).

Supplier (Vendor) Scheduling

8. Percentage delivery releases with full lead time. ☐ ☐
9. Percentage changes to release delivery orders, firm ☐ ☐
schedules, etc.

Material Planning and Material Control

10. Inventory levels (by planner, by product line, by com- ☐ ☐
modity type vs. targets seasonally adjusted, etc.).
11. Safety stock levels (by planner vs. seasonally adjusted ☐ ☐
targets, etc.).
12. Material availability (percentage of items available ☐ ☐
when needed, by week, by planner, by product line,
etc. Percentage of orders able to be released with full
availability, etc.).
13. Planner action/exception messages (percentage of ☐ ☐
items vs. pre-established targets, percentage reviewed
and acted upon where appropriate, etc.).
14. Percentage of schedules/orders released with less than ☐ ☐
planned leadtimes (by planner, by product line, by
vendor, by manufacturing department, etc.).
15. Percentage of schedules/orders rescheduled after ☐ ☐
release (by planner, by product line, by vendor, by
department, etc.).
16. Unplanned activities (material substitutions, product ☐ ☐
design deviations, by planner, by product line, by
department, etc.).

Capacity Planning and Capacity Control

                                                                YES   NO

17. Demonstrated capacity (current levels, as compared to ☐ ☐
    planned capacity, by work center/cell/line).
18. Work-in-Process, Manufacturing Lead Time and ☐ ☐
    Queue levels (by work center/cell/line, in equivalent
    units and/or hours vs. seasonally adjusted targets,
    etc.).
19. Input/Output controls (rates released and completed by ☐ ☐
    work center/cell/line vs. targets, where applicable).
20. Percentage capacity plans past due (by work center/ ☐ ☐
    cell/line, etc.).
21. Percentage of schedules/orders routed to alternate ☐ ☐
    work centers or subcontractors (by product line, by
    department or work center/cell/line).
22. Level of overtime work (by department, work center/ ☐ ☐
    cell/line vs. seasonally adjusted targets).
23. Percentage of operations worked out of sequence (by ☐ ☐
    department, work center/cell/line, etc.).
24. Percentage of schedules/orders split after initial release ☐ ☐
    (by department, work center/cell/line, etc.).
25. Percentage of orders/schedules received on time (by ☐ ☐
    work center/cell/line, by operation, by product line,
    etc.).

Foundation Data

26. Open Order Accuracy (descriptive data, specifications ☐ ☐
    and instructions, timely and accurate progress report-
    ing, etc.).
27. Item Master Accuracy (including planning parameters, ☐ ☐
    descriptive data, etc. by planner, by product, etc.).
28. Routing and Standards Accuracy (specifications, tool- ☐ ☐
    ing, instructions, lead times and run times for set-up,
    manufacturing, move, queue, etc.).
29. Work Center Master Accuracy (including plant capaci- ☐ ☐
    ties, lead times, queue levels, move times, manning
    levels, etc.).

YES   NO

30. Transaction Accuracy (timeliness, variable data accuracy (quantities, etc.), percentage of the time initial transaction input is accepted by system, etc.). □ □

31. Storage Location Accuracy (by line item, by specific location, within defined stocking areas, including point-of-use locations.) □ □

32. Product Change Timeliness (percentage of the time that product design change information is communicated within pre-established timing parameters, hitting pre-established due dates, etc.). □ □

33. Education and Managing Change. Performance against a published plan for initial education should be measured. On-going education needs should be planned and procedures implemented to insure conformance. □ □

Continuous Improvement

34. Each functional area has identified key areas of waste and has active plans to reduce that waste. □ □

35. Each functional area has established a set of key internal or external customer satisfaction measures, tracks performance and seeks the root cause of variations for elimination. □ □

36. Procedures for determining Cost of Quality are documented, data is captured and used to guide continuous improvement activities. □ □

AUDIT QUESTIONS SUMMARY SHEET

I. PLANNING AND CONTROL PROCESSES

A. Strategic Planning   # Yes_____   # No_____
B. Business Planning   # Yes_____   # No_____
C. Sales & Operations Planning   # Yes_____   # No_____
D. Financial Planning, Reporting and
     Measurements   # Yes_____   # No_____
E. "What if?" Simulations   # Yes_____   # No_____
F. Demand Management-Forecasting,
     Sales Planning and Order Entry
     Promising   # Yes_____   # No_____
G. Master Production Scheduling   # Yes_____   # No_____
H. Supplier Planning and Control   # Yes_____   # No_____
I. Material Planning and Control   # Yes_____   # No_____
J. Capacity Planning and Control   # Yes_____   # No_____
K. New/Engineered Product Planning and
     Control   # Yes_____   # No_____
L. Distribution Resource Planning   # Yes_____   # No_____

II. DATA MANAGEMENT

                                                                       # Yes_____   # No_____

III. CONTINUOUS IMPROVEMENT PROCESS

A. Education and Training   # Yes_____   # No_____
B. Continuous Improvement   # Yes_____   # No_____

IV. PERFORMANCE MEASUREMENTS

                                                                       # Yes_____   # No_____

TOTALS   # Yes_____   # No_____

The Mechanics of Manufacturing Resource Planning (MRP II)

Section 1
The Closed Loop System

The basic logic of the closed loop MRP II system is extremely simple. It's in every cookbook. The "bill of material" says, "Turkey stuffing takes one egg, seasoning, bread crumbs, etc." The routing says, "Put the egg and the seasoning in a blender." The blender is the work center. The master schedule is Thanksgiving.

But, in manufacturing, there is a lot more volume and a lot more change. There isn't just one product. There are many. The lead times aren't as short as going to the corner store. The work centers are busy rather than waiting for work—because some of them cost a third of a million dollars or more—and it simply is not wise economically to let them sit idle and to have excess capacity. In addition, the sales department will undoubtedly change the date of Thanksgiving several times before it actually arrives! And this isn't through perversity. This is because the customers want and need some things earlier or later.

The volume of activity in manufacturing is monumentally high; something is happening all the time. And change is the norm, not the exception.

But the point is that the *logic* of MRP II is very straightforward indeed. Figure D-1 shows the closed loop system.

The production plan is the *rate* of production for a product family typically expressed in units like, "We want to produce 1,100 Model 30 pumps per week." The production plan is made by taking into account current inventory, deciding whether inventory needs to go up or down during the planning period, projecting the sales forecast, and determining the rate of production required to maintain, raise, or lower the inventory level. For a make-to-order product, as opposed to a make-to-stock product, the "order backlog" rather than the inventory is the starting point for the production plan.

Figure D-2 shows a typical production plan. Figure D-3 shows a business plan which is simply an extension of the production plan into dollars. The complete business plan in a manufacturing company will include research and development and other expenses not directly related to production and purchases. But the core of any business plan in a manufacturing enterprise is the production plan. With MRP II, the production plan and business plan are interdependent and, as the production plan is updated, it is extended into dollars to show it in the common denominator of business—money.

The MRP II system then takes a master schedule ("What are we going to make?"), "explodes" this through the bill of material ("What does it take to make it?"), and compares

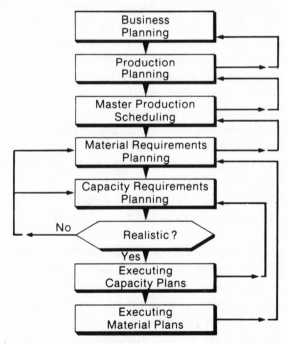

Figure D-1 MRP II

Month Ending		Sales (thousands)	Production (thousands)	Inventory (thousands)
3/31	Plan			
	Actual			60
4/30	Plan	30	35	65
	Actual	25	36	71
6/30	Plan	30	35	75
	Actual			

Figure D-2 Production Plan

this with the inventory on hand and on order ("What do we have?") to determine material requirements ("What do we have to get?").

This fundamental material requirements planning (MRP) logic is shown in Figure D-4. Figure D-5 shows the bill of material. For this example, a small gasoline engine for a moped is the product being manufactured. The bill of material shown in Figure D-5 is what's known as an "indented bill of material." This simply means that the highest level items in the bill of material are shown farthest left. For example, the piston assembly components are "indented" to the right to indicate that they go into that assembly. Therefore, in this example, they are at "level 2."

A bill of material "in reverse" is called a "where-used" list. It would say, for example, that the locating pins go into the crankcase half-left, which goes into the engine.

Month Ending		Sales (thousands)	Production (thousands)	Inventory (thousands)
3/31	Plan			
	Actual			6,000
4/30	Plan	3,000	3,500	6,500
	Actual	2,500	3,600	7,100
5/31	Plan	3,000	3,500	7,000
	Actual	3,800	3,200	6,500
6/30	Plan	3,000	3,500	7,500
	Actual	3,200	3,700	7,000
12/31	Plan	3,000	3,500	10,500
	Actual			

Figure D-3 Business Plan

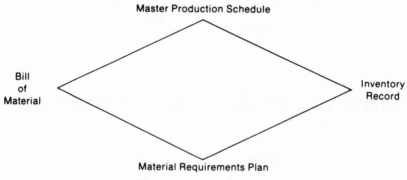

Figure D-4 MRP Logic

```
Part Number
─────────────────
87502    Cylinder Head
94411    Crankshaft
94097    Piston Assembly
  91776      Piston
  84340      Wristpin
  81111      Connecting Rod—Top Half
  27418      Connecting Rod—Bottom Half
  81743      Piston Rings Compression (2)
  96652      Piston Ring Oil
  20418      Bearing Halves (2)
  59263      Lock Bolts (2)
43304    Crankcase Half Right
28079    Crankcase Half Left
  80021      Locater Pins (2)
```

Figure D-5 Moped Engine Bill of Material

Master Production Schedule
Engines

	Week							
	1	2	3	4	5	6	7	8
Master Schedule	80	0	100	0	0	120	0	120
Actual Demand	40	40	30	30	30	40	40	20
Available to Promise	0	0	10	0	0	40	0	100

Figure D-6 Master Production Schedule

Figure D-6 shows a master schedule for engines. In a make-to-stock company, the master schedule would be very similar, but it would take into account the inventory on hand.

Section 2
Material Requirements Planning (MRP)

Figure D-7 shows the material requirements plan for the crankcase half-left and also for the locator pin that goes into the crankcase half-left. The projected gross requirements come from the master schedule plus any service parts requirements. "Scheduled receipts" are the orders that are already in production or out with the vendors. The projected available balance takes the on-hand figure, subtracts requirements from it, and adds scheduled re-

Material Requirements Plan
Crankcase Half — Left

LEAD TIME = 4 WEEKS ORDER QUANTITY = 200	Week							
	1	2	3	4	5	6	7	8
Projected Gross Requirements		80	0	100	0	0	120	0	120
Scheduled Receipts				240					
Proj. Avail. Bal.	120	40	40	180	180	180	60	60	-60
Planned Order Release					200				

Material Requirements Plan
Locater Pin (2 Per)

LEAD TIME = 4 WEEKS ORDER QUANTITY = 500	Week							
	1	2	3	4	5	6	7	8
Projected Gross Requirements					400				400*
Scheduled Receipts									
Proj. Avail. Bal.	430	430	430	430	30	30	30	30	-370
Planned Order Release					500				

*Requirements from Another Crankcase

Figure D-7 Material Requirements Plan

ceipts to it. (In Figure D-7, the starting on-hand balance is 120 for the crankcase half-left.) This calculation projects future inventory balances to indicate when material needs to be ordered or rescheduled.

The material on hand and on order subtracted from the gross requirements yields "net requirements" (60 in week 8 for the crankcase half-left in Figure D-8). This is the amount that is actually needed to cover requirements. When the net requirements are converted to lot sizes and backed off over the lead time, they are called "planned order releases."

The "planned order releases" at one level in the product structure—in this case 200 "crankcase half-left"—become the projected gross requirements at the lower level. The 200-unit planned order release in period four for the crankcase half-left becomes a projected gross requirement of 400 locater pins in period four since there are two locater pins per crankcase half-left.

MRP — Rescheduling
Crankcase Half — Left

LEAD TIME = 4 WEEKS ORDER QUANTITY = 200		Week							
		1	2	3	4	5	6	7	8
Projected Gross Requirements		80	0	100	0	0	120	0	120
Scheduled Receipts					240				
Proj. Avail. Bal.	120	40	40	-60	180	180	60	60	-60
Planned Order Release					200				

MRP — Locater Pin (2 Per)

LEAD TIME = 4 WEEKS ORDER QUANTITY = 500		Week							
		1	2	3	4	5	6	7	8
Projected Gross Requirements					400				400*
Scheduled Receipts									
Proj. Avail. Bal.	430	430	430	430	30	30	30	30	-370
Planned Order Release					500				

*Requirements from Another Crankcase

Figure D-8 MRP—Rescheduling

Most MRP systems also include what is called "pegged requirements." This is simply a way to trace where the requirements came from. For example, the pegged requirements for the locater pins would indicate that the 400 in period four came from the crankcase half-left and that the 400 in period eight came from another product. Pegged requirements show the quantity, the time period, and the higher level item where the requirements are coming from.

Figure D-8 shows the same crankcase half as in Figure D-7. Note, however, that now the scheduled receipt is shown in period four. This means that the due date on the shop order or the purchase order is week four. An MRP system would generate a reschedule message for the planner to move the scheduled receipt from week four into week three to cover the requirements in week three.

Note, also, that the fact that the scheduled receipt for the crankcase half needs to be rescheduled does not affect the requirements for locater pins. The locater pins have already been released into production for the crankcase halves that are on order. The "requirements" for locater pins are for planned orders that have *not* been released yet.

The bill of material is the instrument for converting planned order releases at one level into projected gross requirements at a lower level. The bill of material for the crankcase half-left, for example, would show that two locater pins per crankcase half were required.

Section 3
Capacity Planning and Scheduling

Capacity planning for the manufacturing facility follows the same general logic as the material requirements planning shown in Figure D-4. Figure D-9 shows this capacity requirements planning logic. The remaining operations on released shop orders and all of the operations on planned order releases are "exploded" through the routings (like bills of material for operations) and posted against the work centers (like an inventory of capacities). The result is a capacity requirements plan in standard hours by work center showing the number of standard hours required to meet the material requirements plan. This capacity requirements plan shows the capacity that will be required to execute the master schedule, and consequently, the production plan.

It's important to note that everything in an MRP II system is in "lock step." If the capacity to meet the material requirements plan can't be obtained either through a company's own manufacturing facilities, subcontracting, or purchasing material on the outside, obviously the master schedule will have to be changed. But that is the last resort. The objective is to make the master schedule happen.

Operations scheduling involves assigning individual schedule dates to the operations on a shop order using scheduling rules. Scheduling rules would typically be similar to these:

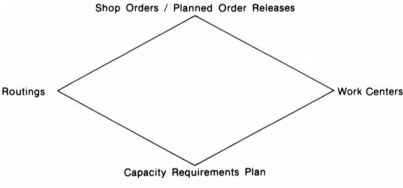

Figure D-9 CRP Logic

1. Allow two days for inspection. (This is a matter of judgment.)

2. Round the standard hours up to the nearest day.

3. Allow X days for queue time.

4. Release work to stockroom one week prior to first operation.

Scheduling with a regular calendar is extremely awkward. For example, if a job was to be completed on August 31 (see Figure D-10) and the last operation—inspection—was scheduled to take two days, the previous operation would have to be completed on August 27, not August 29 (Sunday) or August 28 (Saturday). The scheduler would have to reference the calendar continuously to avoid scheduling work on weekends, holidays, during plant vacation shutdown week, etc. Figure D-11 shows a "shop calendar" where only the working days are numbered. This allows the scheduler to do simple arithmetic like "subtract two days from day 412"; thus the previous operation is to be completed on day 410.

Calendar						
AUGUST						
S	M	T	W	T	F	S
1	2	3	4	5	6	7
8	9	10	11	12	13	14
15	16	17	18	19	20	21
22	23	24	25	26	27	28
29	30	31				

Figure D-10 Calendar

Calendar						
AUGUST						
S	**M**	**T**	**W**	**T**	**F**	**S**
1	2 391	3 392	4 393	5 394	6 395	7
8	9 396	10 397	11 398	12 399	13 400	14
15	16 401	17 402	18 403	19 404	20 405	21
22	23 406	24 407	25 408	26 409	27 410	28
29	30 411	31 412				

Figure D-11 Calendar

Shop calendars are in very common use in manufacturing companies today, but they do have drawbacks. People don't relate to these calendars as easily as they do to a regular calendar. And, of course, they are awkward in dealing with customers who don't use the same shop calendar. Therefore, the shop calendar dates must, once again, be translated back to regular calendar dates. There is a simple solution to this problem with today's computers. A shop calendar can be put in the computer and the computer can do the scheduling using the shop calendar, but print the schedule dates out in regular calendar days. If a company has a shop calendar, there is no reason to discontinue using it if people are used to it. On the other hand, there is no need to introduce the shop calendar today when the computer can do the conversion.

Figure D-12 shows a shop order for the locater pin. This will be used as an example of operations scheduling and, in this example, a shop calendar *will* be used in order to make the arithmetic of scheduling clear. The due date is day 412 and that is determined, in the case of the locater pin that goes into the crank-case half-left, from the material requirements plan.

Operations scheduling works back from this need date to put scheduled finish dates on each operation using scheduling rules like those discussed above. Inspection will be allowed two days. Thus, finish turn must be completed on day 410. It is assumed that the work center file indicates that there are two shifts working in work center 1204 (two shifts at 8 hours apiece equals 16 hours); thus the 27.3 hours required for finish turn will take two days. Planned queue time in this example is assumed to be two days ahead of finish turn. Rough turn must be completed four days earlier than the finish turn must be completed, and its scheduled finish date, therefore, is day 406. The standard hours are calculated by multiplying the quantity by the time per piece and, in this case, adding in the setup time. Where machine operators do not set up their own machines, it might make sense to keep this separate.

It is important to recognize that Figure D-12 shows the information that would be in the computer. *The finish dates would not appear on the shop paperwork that was released to the factory.* The reason is that material requirements planning would be constantly reviewing the need date to see if it had changed. If, for example, the left crankcase halves are scrapped because of a problem with the castings, and the best possible date to have a new lot of castings for the crankcase halves is day 422, the master

Shop Order NN. 18447
Part No. 80021 — Locater Pin
Quant. 500   Due: 412   Release 395

Oper.	Dept.	Work Center	Desc.	Setup	Per Piece	Std. Hrs.	Finish
10	08	1322	Cut Off	.5	.010	5.5	402
20	32	1600	Rough Turn	1.5	.030	16.5	406
30	32	1204	Finish Turn	3.3	.048	27.3	410
40	11		Inspect				412

Figure D-12 Shop Order NN. 18447

schedule would be changed to indicate that. The shop order for the locater pins in the computer would be given a new finish date of 422 and operation 30 would then become 420, operation 20 would become 416, etc.

Capacity requirements will not be posted against the work centers using the routine shown in Figure D-9. A capacity plan, as shown in Figure D-13, will be the result.

This capacity plan has, of course, been cut apart to show it in the figure. It would include many more shop orders, as well as the planned order releases from MRP, in reality. The locater pins are shown here as a released shop order. (Note: there is no released shop order for locater pins in Figure D-8. It would show as a "scheduled receipt" if there were.) One of the great values of MRP is the fact that it projects "planned order releases." These planned order releases are used to:

1. Generate lower level material requirements.

2. Generate capacity requirements.

3. Determine when lower level material—both purchased and manufactured—must be rescheduled to earlier or later dates.

This ability to see capacity requirements ahead of time is especially important to good manpower planning. Seeing the capacity requirements coming rather than seeing the backlogs of

Work Center   1600

Part No.	SO No.	Qty.	Week 396-400	Week 401-405	Week 406-410	Week 411-415	Week 416-420
91762	17621	50		3.5			
80021	18447	500			16.5		

Includes Planned Orders

Part No.	SO No.	Qty.	Week 396-400	Week 401-405	Week 406-410	Week 411-415	Week 416-420
Total Std. Hrs.			294	201	345	210	286

Figure D-13 Capacity Requirements Plan

work out on the factory floor enables factory supervision to do a far better job of leveling production, resulting in less overtime, and less need to hire and lay off people on a short-term basis.

Figure D-14 shows a summary of the capacity requirements over an eight-week period. In practice, this would typically be projected over a far longer period. The summary is drawn from the capacity requirements plan illustrated in Figure D-13 which would also extend much further into the future than the five weeks shown. A typical manpower plan would extend three to six months into the future and would be calculated weekly. A "facilities plan" that would be used for determining what new machine tools were needed would be calculated typically once every two to three months and extended three to four years into the future because of the lead time for procuring machine tools.

The most important information for a foreman is the average hours that he must plan to turn out. This production rate is usually calculated as a four-week average because the individual weekly hours are not particularly significant. The variations between these hours are more random than real. Figure D-13 shows one reason why this happens. The 16.5 hours for part number 80021, the locator pin, are shown in the week bracketed by days 406 to 410. Referring back to Figure D-12, it can be seen that these 16.5 hours are *actually going to be in work center 1600 Tuesday of the previous week!*

Many people have tried to develop elaborate computer load

**Capacity Requirements
Summary (In Standard Hours)**

Week	4-Week Total	4-Week Average	Hours	Week	4-Week Total	4-Week Average	Hours
1	294			5	286		
2	201			6	250		
3	345			7	315		
4	210	1050	263	8	257	1108	277

Figure D-14 Capacity Requirements Summary (in Standard Hours)

leveling systems because they were alarmed by the weekly variation in the apparent "load" shown in the capacity requirements plan. These variations are random. They are exaggerated by the fact that capacity plans are usually done in weekly time periods, and any foreman can attest to the fact that the hours never materialize exactly the same way they are shown on the plan. The most important thing to know is the average rate of output required so that *manpower* can be planned accordingly.

In Figure D-14, the four-week averages are 263 standard hours for the first four weeks and 277 for the second four weeks, or an average of 270 standard hours per week. Now the capacity planner must determine whether that capacity requirement can be met. The first step is to find out what the output from the work center has been over the last few weeks. This is called "demonstrated capacity." (This term was coined by David Garwood and is very useful in describing the present capacity of a work center as opposed to its potential capacity when all shifts are manned, etc.)

It is the job of the capacity planner to then determine whether or not the current capacity is sufficient. Or, what needs to be done to get the capacity to meet the plan. Or—as a last resort—to feed back information that the plan cannot be met.

If the plan cannot be met, the master schedule and, perhaps, even the production plans will have to be changed. If, for example, a company has one broach and it is the only one of its type available because it was made specifically for this company, it could well become a bottleneck. If the capacity plan indicates that more hours were required at the broach than could possibly be produced, the master schedule would have to be changed to reflect this.

Once again, however, it's important to emphasize that this is the *last resort*. The job of the capacity planner is to get the capacity that is needed to meet the plan. And that is an important point to emphasize. If there is any problem that exists in practice with capacity planning, it is the fact that people expect the computer to do the capacity planning rather than recognizing that all it can do is generate numbers that will be given to an intelligent, experienced person—the capacity planner—to use in working with other people to fix capacity problems.

Once it is agreed that the capacity requirements can be met, an output control report as shown in Figure D-15 is set up. Three weeks have passed since the one in the figure was made, and the actual standard hours produced (shown in the second line of the figure) are falling far short of the required standard hours at work center 1600. The deviation in the first week was 20 hours. In the second week, it was 50 hours—for a cumulative deviation of 70 hours. In the third week, it was 80 hours, giving a total cumulative deviation of 150 hours. This is a true *control* report with a plan and feedback to show where actual output in standard hours compares with the plan. It shows the deviation from the plan. The 150 hour deviation in week three indicates that 150 standard hours of work required to produce material to meet the master schedule has not been completed.

The amount of tolerance around the plan has to be established. If it were determined, for example, that the company could tolerate being one half week behind schedule, the tolerance in Figure D-15 would be 135 standard hours. When the deviation exceeds 135 standard hours, that would require immediate attention to increase output through overtime, adding people, etc. Whenever the planned rate in the output control report is changed, the deviation will be reset to 0.

It's a good idea to show input to a work center as well as output. That way, when a work center is behind on output because a feeding work center has not given them the work, it can be de-

Output Control
Work Center 1600
Week No. 4
(in Std. Hrs.)

Today

	Week 1	Week 2	Week 3	Week 4
Planned	270	270	270	270
Actual Std.	250	220	190	
Deviation	-20	-70	-150	

Figure D-15 Output Control

tected very quickly since the input report will show the actual input below the planned input. This is called an "input/output report."

The capacity planning and output control reports are concerned with capacity. The dispatch list shown in Figure D-16 is concerned with priority.

The dispatch list is generated daily—or as required—and goes out to the shop floor at the beginning of the day. It shows the sequence in which the jobs are to be run according to the scheduled date for the operation in that work center. The movement of jobs from work center to work center is put in to the computer so that each morning the foremen can have an up-to-date schedule that is driven by MRP. If part 80021 had been rescheduled to a new completion date of day 422 as discussed above, its priority would drop on the dispatch list because its scheduled date would now be 416. This would allow part number 44318 to be made earlier. The dispatch list gives the foremen the priority of jobs so that they can pick the proper job to start next. Since the dispatch list is driven by MRP, it tells the foremen the right sequence in which to run the jobs to do the best job of preventing predicted shortages.

Dispatch List				Day 405
Work Center No. 1600				
Shop Order No.	Part No.	Qty.	Scheduled Date	Std. Hours
17621	91762	50	401	3.5
18430	98340	500	405	19.2
18707	78212	1100	405	28.6
18447	80021	500	406	16.5
19712	44318	120	409	8.4
			Total Hours	**76.2**

Figure D-16 Dispatch List

Section 4
The MRP II Output Reports

The figures thus far in this appendix represent the major operating reports that are used in an MRP II system. Referring back to Figure D-1, the functions of the production plan (Figure D-2), the master schedule (Figure D-6), the material requirements plan (Figures D-7 and D-8), and the capacity requirements plan (Figure D-13) are illustrated. The output control report (Figure D-15) is the means for monitoring output against the plan to be sure that capacity plans are being executed. The dispatch list (Figure D-16) is the report for the factory to use in executing the material plans. Vendor scheduling is the way the material requirements plans are executed with the "outside factories."

It is important to emphasize the feedback functions in a closed loop system. For example, if vendors are not going to ship on time, they must send in an anticipated delay report as soon as they recognize that they have a problem. In the past, ship dates were not valid. The typical company had many past due purchase orders with the vendor. With MRP II—if it is properly managed— dates will represent real need dates, and, thus, it is important to feed back information as quickly as possible to indicate when these dates cannot be met. This, of course, is also true for the factory, where the anticipated delay report should be a regular part of their feedback to the closed loop system.

The financial reports that can be a by-product of these operating reports in an MRP II system are shown in Figure D-17.

We've already discussed the relationship of the production plan and business plan (see Figures D-1 and D-2). In a company where the production plans are kept current and costed out properly, they should be the basis for the business plan. Actual sales, production, and inventory can be recorded against the production plans and they can be used as *control reports.* The business plan can be kept up to date as production plans are changed. Management can see the financial impact of changes in the production plans on the business plan.

The master schedule, costed out, is the basis for *"transfers to*

Operations	Finance
Production Plans	Business Plan
Master Schedules	Shipping budget (make-to-stock companies), transfers to inventory (make-to-stock companies)
Material Requirements Plans	Current inventories, projected consumption, purchase commitments, manufacturing schedules, and projected future inventory balances
Capacity Requirements Plans	Labor requirements by labor grade
Input/Output Reports	Standard hours of output by work center in units and dollars
Dispatch Lists	Work in process, labor reporting, efficiency reports
Vendor Schedules	Commitments by vendor

Figure D-17 Financial Reports﹀

inventory" in a make-to-stock company or the *shipping budget* in a make-to-order company. This is, of course, the same as "production" in the production plan. Management can review the projected shipping budgets to make sure that the objectives of shipping to budget and shipping the *right* orders to the *right* customers are being properly reconciled.

Figure D-18 shows the inventory of components for a company making pumps. The inventory file is coded to show which components go into the pumps. The first column in the projected available balance shows the on-hand inventory in units extended by the cost. This is the current number of dollars of pump components in inventory. Projected gross requirements from MRP are costed out, as are scheduled receipts. The projected available balance shows the projected future stockroom inventory month by month based on the projected gross requirements derived from the master schedule which, in turn, is derived from the production plan and, in an MRP II system, represents the detailed execution of the business plan.

Costing out the material requirements plan and summarizing it by product group categories results in:

1. On-hand inventory (in dollars) by product group.

2. How much material will be consumed (in dollars) to support the current production plans ("requirements").

Pump Component Inventory — in $(000)

		Month			
		1	2	3	4
Projected Gross Requirements		250	250	300	300
Scheduled Receipts		250	250	270	290
Proj. Avail. Bal.	540	540	540	510	500

Figure D-18

3. What will have to be purchased to support the current production plans (in dollars).

4. What the projected stockroom inventory balances by product group in dollars should be for months into the future.

5. What will have to be made—this is the shop schedule that will become input to capacity planning and can be converted into labor dollars.

Comparing the actual withdrawals from stock against "requirements," the actual purchases against the plan, and the actual inventory dollars against the projected balances, tells management if the plans are really being executed—and if the *business plan* is actually being executed at the detail level.

From a manufacturing point of view, the output of material requirements planning is released shop orders and planned order releases—the shop schedule. When this is run through the capacity requirements planning section of an MRP II system, the result is the standard hours by work center, by time period, required to satisfy the master schedule and the production plans. This can be converted to labor dollars by labor grade, by time period, and by product group. From this information, management can see how many dollars of material and labor will have to be purchased by time period to support any given production plan with its particular product mix. Some companies, in order to project cash flow with greater accuracy, actually offset the due dates by the payable dates.

Costing out the open shop order file that is used to make the dispatch list yields current work in process in dollars. Labor reporting is usually tied in with dispatching and is the basis for labor efficiency reports. The vendor schedules costed out tell how much of the purchased material is due to be shipped *by vendor* by time period to support the plan.

Perhaps, the most important result of an operating system that can work is inventory valuation. When the inventory records are correct—as they have to be for an MRP system to operate—having accounting cost these records out to get the value of the inventory is a very straightforward matter.

When the operating system works, using it to drive the accounting system means that accounting has better numbers to work with than ever before. It also means that the operating system makes more sense to management than ever before because dollars are the language of business. And when the operating system and the financial system are saying the same things, the financial people and the operating people can talk the same language to management rather than presenting conflicting information.

(Note: Material in Appendix A is taken from *Manufacturing Resource Planning: MRP II—Unlocking America's Productivity Potential* by Oliver Wight, Oliver Wight Limited Publications, Inc., 1984)

Index